Gabriel Jacques

Trophic interaction between predatory bedbugs

Gabriel Jacques

Trophic interaction between predatory bedbugs

And their importance in biological pest control

ScienciaScripts

Imprint
Any brand names and product names mentioned in this book are subject to trademark, brand or patent protection and are trademarks or registered trademarks of their respective holders. The use of brand names, product names, common names, trade names, product descriptions etc. even without a particular marking in this work is in no way to be construed to mean that such names may be regarded as unrestricted in respect of trademark and brand protection legislation and could thus be used by anyone.

Cover image: www.ingimage.com

This book is a translation from the original published under ISBN 978-3-330-75515-4.

Publisher:
Sciencia Scripts
is a trademark of
Dodo Books Indian Ocean Ltd. and OmniScriptum S.R.L publishing group

120 High Road, East Finchley, London, N2 9ED, United Kingdom
Str. Armeneasca 28/1, office 1, Chisinau MD-2012, Republic of Moldova, Europe
Printed at: see last page
ISBN: 978-620-8-27090-2

CONTENTS

INTRODUCTION 2
REFERENCES 7
CHAPTER I 15
CHAPTER II 39
CONSIDERATIONSFINAL 68

INTRODUCTION

The use of biological control

Insect pests can cause losses in agricultural and forestry production. Chemical control of these species is efficient, but can limit populations of natural enemies, cause poisoning, environmental contamination, increase costs and lead to the emergence of resistant insects (Zanuncio *et al.*, 1994). This motivates the search for control methods to reduce pest population densities, especially biological control, which is important in integrated pest management (Zanuncio *et al.*, 1998a).

Biological control is the use of living organisms to suppress populations or the impact of pests. Biological control agents are mainly used for: (1) biological control of pest invertebrates with predators, parasitoids and pathogens, (2) biological control of weeds with pathogens and herbivores, and (3) biological control of plant pathogens with antagonistic microorganisms and induction of plant resistance (Eilenberg *et al,* 2001).

Predatory heteropterans are used to control pests. In Europe, *Podisus maculiventris* (Say) (Heteroptera: Pentatomidae) has been used since 1997 against defoliating caterpillars of vegetables and ornamental plants in greenhouses, such as *Spodoptera exigua* (Hübner) and *Chrysodeixis chalcites* (Esper) (Lepidoptera: Noctuidae) (De Clercq *et al.*, 1998). In the United States, *Geocoris punctipes* (Say) (Heteroptera: Lygaeidae) and *Orius tristicolor* (Say) (Heteroptera: Anthocoridae) showed a high predation capacity on adults of the whitefly *Bemisa tabaci* (Gennadius) (Hemiptera: Aleyrodidae) (Hagler *et al.*, 2004). In Brazil, *Podisus nigrispinus* (Dallas), *Podisus distinctus* (Stâl), *Brontocoris tabidus* (Signoret) and *Supputius cincticeps* (Stâl) (Heteroptera: Pentatomidae) are used in integrated pest management in eucalyptus plantations (Zanuncio *et al.*, 1994).

Podisus nigrispinus is a generalist and Neotropical predator (Lemos *et al*, 2005). In Brazil, this species is important in the biological control (Pereira *et al.*, 2008; Ramalho *et al.*, 2008) of defoliator caterpillars (Zanuncio *et al.*, 1994). Forestry companies in the state of Minas Gerais, Brazil, have studied and used *P. nigrispinus* to protect eucalyptus plantations against *Thyrinteina arnobia* (Stoll) (Geometridae), *Psorocampa denticulata* (Schaus) (Notodontidae), *Eupseudosoma aberrans* Schaus (Arctiidae), *Sarsina violascens* (Herrich-Schaeffer) (Lymantriidae) and other defoliating lepidopterans. More than five million of these predators have been released, with significant reductions in insecticide use (Neves *et al.*, 2009). In addition, this species has potential for biological control programs of *Anticarsia gemmatalis* Hübner (Noctuidae), in soybean plantations (Ferreira *et al.*, 2008) and *Alabama argillacea* (Hübner) (Noctuidae), in cotton plantations (Pereira *et al.*, 2009) in Brazil.

Supputius cincticeps is also a generalist of the Neotropical region and has been reported as a predator of agricultural and forestry pests, mainly in South America, with potential for the biological control of *Spodoptera frugiperda* (Smith) (Lepidoptera: Noctuidae) in corn (Zanuncio *et al., 1998b*), 1998b); *T. arnobia* in eucalyptus (Zanuncio *et al.*, 2001a) and *A. gemmatalis* in soybeans (Corso *et al.*, 1999).

Podisus nigrispinus and *S. cincticeps* can be affected by other natural enemies, which can compromise their efficiency in biological control.

Increased diversity of natural enemies

Interactions between natural enemies and the influence of these organisms on community structure and the effectiveness of biological control need to be studied. The use of a greater diversity of natural enemies is beneficial in pest control, as it can increase the reduction of herbivore populations and the more diverse they are, the longer the time without disturbing ecosystems (Aquilino *et al.*, 2005; Byrnes *et al.*, 2005; Wilby *et al.*, 2005; Snyder *et al.*, 2006, 2008; Herrick *et al.*, 2008; Letourneau *et al.*, 2009; Northfield *et al.*, 2010). However, in some cases, the addition of more than one natural enemy can reduce the efficiency of biological control (Snyder and Wise, 2001; Finke and Denno, 2004, 2005; Bruno and O'Connor, 2005).

Prey can be consumed more or less by biological control agents (Casula *et al.* 2006; Janssen *et al.*, 2007) and present three types of interspecific interactions, (1) synergism, with increased mortality of the prey in relation to each natural enemy or independently, when the foraging behavior of one species facilitates the capture of the pest by other natural enemies. *Harmonia axyridis* (Pallas) (Coleoptera: Coccinellidae), *Nabis* sp. (Latreille) (Hemiptera: Nabidae) and *Aphidius ervi* (Haliday) (Hymenoptera: Braconidae) suppressed populations of *Acyrthosiphon pisum* (Harris) (Homoptera: Aphididae) beyond what was expected for each natural enemy (Cardinale *et al.*, 2003). The combined use of natural enemies can provide better control of the pest (Bjorkman and Liman, 2005). (2) The total mortality of the pest by natural enemies together equals the sum of the individual mortality by each one. This occurs for independent natural enemies, when they prey on different stages of the pest's life or at different times. *Dicyphus tamaninii* Wagner and *Macrolophus caliginosus* (Wagner) (Heteroptera: Miridae) did not show negative interactions in the biological control of the whitefly *Bemisia tabaci* Gennadius (Homoptera: Aleyrodidae), but without increasing the predation rate (Lucas and Alomar, 2002). Even without synergy, these predators are recommended together. (3) The total mortality of the pest by predators together is lower than the sum of the individual mortality per predator. The decrease in the predation rate can occur through different mechanisms, including intraguild predation and interspecific competition (Huxel, 2007). *Zelus renardii* Kolenati (Heteroptera: Reduviidae) and *Nabis* sp. (Latreille) (Heteroptera: Nabidae) increased the mortality of larvae of *Chrysoperla carnea* (Stephens) (Neuroptera: Chrysopidae), a predator of *Aphis gossypii* Glover (Hemiptera: Aphididae) (Rosenheim *et al.*, 1993), which reduced the efficiency of biological control.

Intra-guild predation and competition

Intraguild predation occurs when a species competes for prey and also consumes its competitor (Pell *et al.*, 2008). The abundance of extraguild prey and the feeding specificity of each species affects the likelihood of intraguild prey occurring, while this interaction is often determined by the mobility, size and/or developmental stage of the predators and the prey (Phoofolo and Obrycki, 1998; Lucas, 2005). A decrease in the number of extraguild prey increases intraguild predation (Sato *et al.*, 2009). Intraguild predation by the Asian ladybug, *Harmonia axyridis* (Pallas) (Coleoptera: Coccinellidae), can reduce populations of other coccinellids through intraguild predation (Koch, 2003; Cottrell, 2005; Koch *et al.*, 2006).

The importance of intraguild predation in biological control is still unclear, with disagreements as to whether the addition of an intraguild predator increases or decreases pest suppression (Vance-Chalcraft *et al.,* 2007). On the one hand, intraguild predation can exclude intraguild prey and/or reduce time and effort in preying on the target pest (Rosenheim *et al.*, 1995; Snyder *et al.*, 2008). This behavior can reduce top-down effects on herbivores, weaken trophic cascades and impair biological control (Denno *et al.*, 2004; Finke and Denno, 2004, 2005). On the other hand, intraguild predation can help maintain predator populations when extraguild prey is scarce, and is therefore beneficial to biological control (Ruberson *et al.*, 1986). In addition, the prey's behavior of avoiding more than one predator leads to sublethal consequences for the prey, such as reduced feeding and reproduction time, which can lead to lower population growth (Nelson, 2007).

Behavioral adaptations that reduce the probability of encounters or the chemical or physical defense of each predator reduce intraguild predation (Sato and Dixon, 2004; Urbani and Ramos-Jiliberto, 2010). The intraguild prey may select a preferred habitat to avoid interactions with the intraguild predator, although this may involve a trade-off of access to high-quality resources (Heithaus, 2001). This selection can be made by clues released by the intraguild predator, such as feces (Agarwala *et al*, 2003), semiochemicals and trails (Persons and Rypstra, 2001; Nakashima *et al.*, 2006). However, when resources for intraguild prey become scarce, they may leave their safe habitat in search of new resources, increasing the chances of encountering the intraguild predator (Creswell *et* al., 2010; Wilson *et al.*, 2010). Alternative resources for intraguild prey also reduce intraguild predation (Holt and Huxel, 2007; Okuyama, 2009). These alternative resources can be co-specific prey (Rudolf and Armstrong, 2008). Interaction is reduced when species forage in the same environment at different times (Snyder *et al.,* 2008), seasons (Sato and Dixon, 2004) or microhabitats (Janssen *et al.*, 2007). *Coccinella septempunctata brucki* Mulsant (Coleoptera: Coccinellidae) can be preyed upon by *H. axyridis*, but lays its eggs earlier in the spring, i.e. *H. axyridis* rarely encounters eggs and young larvae of *C. Septempunctata* (Sato and Dixon, 2004). Intraguild prey are less profitable for many predators, which is defined as the energy content absorbed divided by the time spent handling the prey, compared to extraguild prey (Walzer and Schausberger, 2011). This is because intraguild prey are also predators and can counter-attack the intraguild predator, so the energy cost of intraguild predation can outweigh the benefits (Lawson-Balagbo *et al.*, 2008). In addition, intraguild prey are generally less abundant than extraguild prey, reducing the likelihood of intraguild predation (Sato *et al.,* 2009).

Different species of natural enemies attracted to plants can interact through competition. Interspecific competition occurs when species from the same guild compete for limited resources, which can result in the survival of the better adapted species and a decrease in the population of the inferior competitor (Ware *et al*., 2009). The magnitude of this competition can vary due to the complexity of the environment and the effects of priority, which refers to the sequence in which competitors arrive in an environment (Bonin *et al*., 2009; Geange and Stier, 2010). More complex habitats generally contain a greater diversity of food resources and refuges for predators, reducing the intensity of intraguild competition and predation (Finke and Denno, 2002; Juliano, 2009). Arriving first in an environment can confer advantages due to greater adaptation to the local environment, which facilitates the ability to acquire resources and escape from predators (Geange and Stier, 2009).

The coexistence of competitors occurs through the joint occurrence of: (1) spatial and/or temporal heterogeneity of the resource (Schmidt *et al*., 2000) or the availability of different life stages of the prey (Wilby and Thomas, 2002; Gonçalves *et al., 2006) and (2)* different life histories or foraging adaptations of the competitors in relation to the availability of resources (Schmidt *et al.,* 2000; Moser and Obrycki, 2009). The predators *Typhlodromalus aripo* De Leon and *Typhlodromalus manihoti* (Moraes) (Acari: Phytoseiidae) prey on the cassava mite *Mononychellus tanajoa* (Bondar) (Acari: Tetranychidae) on different parts of the plant, with *T. aripo* on the apex and *T. manihoti* on the leaves. In addition, at low prey densities, *T. aripo* outnumbers *T.*

manihoti; and at high densities the reverse occurs (Magalhaes *et al.*, 2003). Each predator species has its own life history, with characteristics that allow it to successfully exploit the necessary prey densities in its microhabitats.

Podisus nigrispinus and *S cincticeps* are used in biological control and are found in the same habitat, but the effects of competition and possible intraguild predation between these predators have not been studied. Biological characteristics suitable for pest control, such as high predation and reproduction capacity, easy rearing and low cost (Zanuncio *et al.*, 1994) make it possible to use these natural enemies simultaneously, however, both attack common prey and this makes it necessary to study the interactions between them in biological control programs.

REFERENCES

Agarwala, B.K., Yasuda, H. and Kajita, Y. 2003. Effect of conspecific and heterospecific feces on foraging and oviposition of two predatory ladybirds: role of fecal cues in predator avoidance. Journal of Chemical Ecology 29, 357-376.

Aquilino, K.M., Cardinale, B.J., Ives, A.R. 2005. Reciprocal effects of host plant and natural enemy diversity on herbivore suppression: an empirical study of a model tritrophic system. Oikos 108, 275-282.

Bjorkman, C. and Liman, A.S. 2005. Foraging behavior influences the outcome of predator-predator interactions. Ecological Entomology 30, 164-169.

Bonin, M.C., Srinivasan, M., Almany, G.R. and Jones, G.P. 2009. Interactive effects of interspecific competition and microhabitat on early postsettlement survival in a coral reef fish. Coral Reefs 28, 265-274.

Bruno, J.F. and O'Connor, M.I. 2005. Cascading effects of predator diversity and omnivory in a marine food web. Ecology Letters 8, 1048-1056.

Byrnes, J., Stachowicz, J.J., Hultgren, K.M., Hughes, A.R., Olyarnik, S.V. and Thornbert, C.S. 2005. Predator diversity strengthens trophic cascades in kelp forests by modifying herbivore behavior. Ecology Letters 9, 61-71.

Cardinale, B.J., Harvey, C.T., Gross, K. and Ives, A.R. 2003. Biodiversity and biocontrol: emergent impacts of a multiple-enemy assemblage on pest suppression and crop yield in an agroecosystem. Ecology Letters 6, 857-865.

Casula, P., Wilby A. and Thomas M.B. 2006. Understanding biodiversity effects on prey in multi-enemy systems. Ecology Letters 9, 995-1004.

Cottrell, T.E. 2005. Predation and cannibalism of lady beetle eggs by adult lady beetles. Biological Control 34, 159-164.

Corso, I.C., Gazzoni, D.L., Nery, M.E. 1999. Effect of doses and refuge on the selectivity of insecticides to predators and parasitoids of soybean pests. Pesquisa Agropecuària Brasileira 34, 1529-1538.

Creswell, W., Lind, J. and Quinn, J.L. 2010. Predator-hunting success and prey vulnerability: quantifying the spatial scale over which lethal and non-lethal effects of predation occur. Journal of Animal Ecology 79, 556-562.

De Clercq, P., Merlevede, F., Mestdagh, I., Vandendurpel, K., Mohaghegh J. and Degheele, D. 1998. Predation on the tomato looper *Chrysodeixis chalcites* (Esper) (Lep., Noctuidae) by *Podisus maculiventris* (Say) and *Podisus nigrispinus* (Dallas) (Het., Pentatomidae). Journal of Applied Ecology 122, 93-98.

Denno, R.F., Mitter, M.S., Langellotto, G.A., Gratton, C. and Finke, D.L. 2004. Interactions between a hunting spider and a web-builder: consequences of intraguild predation and cannibalism for prey suppression. Ecological Entomology 29, 566-577.

Eilenberg, J., Hajek, A. and Lomer, C. 2001. Suggestions for unifying the terminology in biological control. BioControl 46, 387-400.

Ferreira, J.A.M., Zanuncio, J.C., Torres, J.B., Molina-Rugama, A.J. 2008. Predatory behavior of *Podisus nigrispinus* (Heteroptera : Pentatomidae) on different densities of *Anticarsia gemmatalis* (Lepidoptera : Noctuidae) larvae. Biocontrol Science and Technology 18(7), 711-719.

Finke, D.L. and Denno, R.F. 2002. Intraguild predation diminished in complex-structured vegetation: implications for prey suppression. Ecology 83, 643652.

Finke, D.L. and Denno, R.F. 2004. Predator diversity dampens trophic cascades. Nature 429, 407-410.

Finke, D.L. and Denno, R.F. 2005. Predator diversity and the functioning of ecosystems: the role of intraguild predation in dampening trophic cascades. Ecology Letters 8, 1299-1306.

Geange, S.W. and Stier, A.C. 2010. Priority effects and habitat complexity affect the strength of competition. Oecologia 163, 111-118.

Geange, S.W. and Stier, A.C. 2009. Order of arrival affects competition in two reef fishes. Ecology 90, 2868-2878.

Gonçalves, J.R., Faroni L.R.D., Guedes R.N.C., Oliveira, C.R.F. and Garcia, F.M. 2006. Interaction between *Acarophenax lacunatus* (Cross & Krantz) (Prostigmata: Acarophenacidae) and *Anisopteromalus calandrae* (Howard)

(Hymenoptera: Pteromalidae) on *Rhyzopertha dominica* (Fabricius) (Coleoptera: Bostrichidae). Neotropical Entomology 35, 823-82.

Hagler, J.R., Jackson, C.G., Isaacs, R. and Machtley, S.A. 2004. Foraging behavior and prey interactions by a guild of predators on various lifestages of *Bemisa tabaci*. Journal of Insect Science 4, 1-13.

Heithaus, M.R. 2001. Habitat selection by predators and prey in communities with asymmetrical intraguild predation. Oikos 92, 542-554.

Herrick, N.J., Reitz, S.R., Carpenter, J.E. and O'Brien, C.W. 2008. Predation by *Podisus maculiventris* (Hemiptera: Pentatomidae) on *Plutella xylostella* (Lepidoptera: Plutellidae) larvae parasitized by *Cotesia plutellae* (Hymenoptera: Braconidae) and its impact on cabbage. Biological Control 45, 386-395.

Holt, R.D. and Huxel, G.R. 2007. Alternative prey and the dynamics of intraguild predation: theoretical perspectives. Ecology 88, 2706-2712.

Huxel, G.R. 2007. Antagonistic and synergistic interactions among predators. Bulletin of Mathematical Biology 69, 2093-2104.

Janssen, A., Sabelis M.W., Magalhaes S., Montserrat M. and van der Hammen T. 2007. Habitat structure affects intraguild predation. Ecology 88, 2713-2719.

Juliano, S.A. 2009. Species interactions among larval mosquitoes: context dependence across habitat gradients. Annual Review of Entomology 54, 37-56.

Koch, R.L. 2003. The multicolored Asian lady beetle, *Harmonia axyridis*: a review of its biology, uses in biological control, and non-target impacts. Journal of Insect Science 3, 1-16.

Koch, R.L., Venette, R.C. and Hutchison, W.D. 2006. Invasions by *Harmonia axyridis* (Pallas) (Coleoptera: Coccinellidae) in the Western Hemisphere: implications for South America. Neotropical Entomology 35, 421-434.

Lawson-Balagbo, L.M., Gondim, M.G.C.Jr., de Moraes, G.J., Hanna, R. and Schausberger, P. 2008. Compatibility of *Neoseiulus paspalivorus* and *Proctolaelaps bickleyi*, candidate biocontrol agents of the coconut mite *Aceria guerreronis*: spatial niche use and intraguild predation. Experimental and Applied Acarology 45, 1-13.

Lemos, W.P., Ramalho, F.S., Serrao, J.E. and Zanuncio, J.C. 2005. Morphology of female reproductive tract of the predator *Podisus nigrispinus* (Dallas) (Heteroptera: Pentatomidae) fed on different diets. Brazilian Archives of Biology and Technology 48, 129-138.

Letourneau, D.K., Jedlicka J.A., Bothwell S.G. and Moreno, C.R. 2009. Effects of natural enemy biodiversity on the suppression of arthropod herbivores in terrestrial ecosystems. The Annual Review of Ecology, Evolution and Systematics 40, 573-92.

Lucas, E. 2005. Intraguild predation among aphidophagous predators. European Journal of Entomology 102, 351-364.

Lucas, E. and Alomar, O. 2002. Impact of the presence of *Dicyphus tamaninii* Wagner (Heteroptera: Miridae) on whitefly (Homoptera: Aleyrodidae) predation by *Macrolophus caliginosus* (Wagner) (Heteroptera: Miridae). Biological Control 25, 123-128.

Magalhaes, S., Brommer, J.E., Silva, E.S., Bakker, F.M. and Sabelis, M.W. 2003. Life-history trade off in two predator species sharing the same prey: a study on cassava-inhabiting mites. Oikos 102, 533-542.

Moser, S.E. and Obrycki, J.J. 2009. Competition and intraguild predation among three species of coccinellids (Coleoptera: Coccinellidae). Annals of the Entomological Society of America 102, 419-425.

Nakashima, Y., Birkett, M.A., Pye, B.J. and Powell, W. 2006. Chemically mediated intraguild predator avoidance by aphid parasitoids: interspecific variability in sensitivity to semiochemical trails of ladybird predators. Journal of Chemical Ecology 32, 1989-1998.

Nelson, E.H. 2007. Predator avoidance behavior in the pea aphid: costs, frequency, and population consequences. Oecologia 151, 22-32.

Neves, R.C.S., Torres, J.B and Vivan, L.M. 2009. Reproduction and dispersal of wing-clipped predatory stinkbugs, *Podisus nigrispinus* in cotton fields. BioControl 54, 9-17.

Neves, R.C.S., Torres, J.B and Zanuncio, J.C. 2010. Production and storage of mealworm beetle as prey for predatory stinkbug. Biocontrol Science and Technology 20, 1013-1025.

Northfield, T.D., Snyder, G.B., Ives, A.R. and Snyder, W.E. 2010. Niche saturation reveals resource partitioning among consumers. Ecology Letters 13, 338348.

Pell, J.K., Baverstock, J., Roy, H.E., Ware, R.L. and Majerus, M.E.N. 2008. Intraguild predation involving *Harmonia axyridis*: a review of current knowledge and future perspectives. BioControl 53, 147-168.

Pereira, A.I.A., Ramalho, F.S., Bandeira, C.M., Malaquias, J.B. and Zanuncio, J.C. 2009. Age-dependent fecundity of *Podisus nigrispinus* (Dallas) (Heteroptera: Pentatomidae) with sublethal doses of gammacyhalothrin. Brazilian Archives of Biology and Technology 52, 1157-1166.

Pereira, A.I.A., Ramalho, F.S., Malaquias, J.B., Bandeira, C.M., Silva, J.P.S. and Zanuncio, J.C. 2008. Density of *Alabama argillacea* larvae affects food extraction by females of *Podisus nigrispinus*. Phytoparasitica 36, 84-94.

Persons, M.H. and Rypstra, A.L. 2001. Wolf spiders show graded antipredator behavior in the presence of chemical cues from different sized predators. Journal of Chemical Ecology 27, 2493-2504.

Phoofolo, M.W. and Obrycki, J.J. 1998. Potential for intraguild predation and competition among predatory Coccinellidae and Chrysopidae. Entomologia Experimentalis et Applicata 89, 47-55.

Okuyama, T. 2009. Intraguild predation in biological control: consideration of multiple resource species. Biocontrol 54, 3-7.

Ramalho, F.S., Mezzomo, J., Lemos, W.P., Bandeira, C.M., Malaquias, J.B., Silva, J.P.S., Leite, G.L.D. and Zanuncio, J.C. 2008. Reproductive strategy of *Podisus nigrispinus* females under different feeding intervals. Phytoparasitica 36, 3037.

Rosenheim, J.A., Wilhoit, L.R. and Armer,C.A. 1993. Non-additive effects of multiple natural enemies on aphid populations. Oecologia 108, 375-379.

Rosenheim, J.A., Kaya, H.K., Ehler, L.E., Marois, J.J. and Jaffee, B.A. 1995. Intraguild predation among biological control agents: theory and evidence. Biological Control 5, 303-335.

Ruberson, J.R., Tauber, M.J. and Tauber, C.A. 1986. Plant feeding by *Podisus maculiventris* (Heteroptera: Pentatomidae): effect on survival, development, and preoviposition period. Environmental Entomology 15, 894-897

Rudolf, V.H.W. and Armstrong, J. 2008. Emergent impacts of cannibalism and size refuges in prey on intraguild predation systems. Oecologia 157, 675-686.

Sato, S., Shinya, K., Yasuda, H., Kindlmann, P. and Dixon, A.F.G. 2009. Effects of intra and interspecific interactions on the survival of two predatory ladybirds (Coleoptera: Coccinellidae) in relation to prey abundance. Applied Entomology and Zoology 44, 215-221.

Sato, S. and Dixon, A.F.G. 2004. Effect of intraguild predation on the survival and development of three species of aphidophagous ladybirds: consequences for invasive species. Agricultural and Forest Entomology 6, 21-24.

Schmidt, K.A., Earnhardt, J.M., Brown, J.S. and Holt, R.D. 2000. Habitat selection under temporal heterogeneity: exorcizing the ghost of competition past. Ecology 81, 2622-2630.

Snyder, W.E. and Wise, D.H. 2001. Contrasting trophic cascades generated by a community of generalist predators. Ecology 82, 1571-1583.

Snyder, G.B., Finke, D.L. and Snyder, W.E. 2008. Predator biodiversity strengthens aphid suppression across single- and multiple-species prey communities. Biological Control 44, 52-60.

Snyder, W.E., Snyder, G.B., Finke, D.L. and Straub, C.S. 2006. Predator biodiversity strengthens herbivore suppression. Ecology Letters 9, 789-796.

Urbani, P. and Ramos-Jiliberto, R. 2010. Adaptive prey behavior and the dynamics of intraguild predation systems. Ecological Modelling 221, 2628-2633.

Vance-Chalcraft, H.D., Rosenheim, J.A., Vonesh, J.R., Osenberg, C.W. and Sih, A. 2007. The influence of intraguild predation on prey suppression and prey release: a meta-analysis. Ecology 88, 2689-2696.

Walzer, A. and Schausberger, P. 2011. Threat-sensitive anti-intraguild predation behavior: maternal strategies to reduce offspring predation risk in mites. Animal Behaviour 81, 177-184.

Ware, E.R., Yguel, B. and Majerus, M. 2009. Effects of competition, cannibalism and intra-guild predation on larval development of the European coccinellid *Adalia bipunctata* and the invasive species *Harmonia axyridis*. Ecological Entomology 34, 12-19.

Wilby, A. and Thomas, M. 2002. Natural enemy diversity and pest control: patterns of pest emergence with agricultural intensification. Ecology Letters 5, 353360.

Wilby, A., Villareal, S.C., Lan, L.P., Heong, K.L. and Thomas, M.B. 2005. Functional benefits of predator species diversity depend on prey identity. Ecological Entomology 30, 497-501.

Wilson, R.R, Blankenship, T.L., Hooten, M.B. and Shivik, J.A. 2010. Prey- mediated avoidance of an intraguild predator by its intraguild prey. Oecologia 164, 921-929.

Zanuncio, J.C., Alves, J.B., Zanuncio, T.V. and Garcia, J.F. 1994. Hemipterous predators of eucalypt defoliator caterpillars. Forest Ecology and Management 65, 6573.

Zanuncio, J.C., Mezzono, J.A., Guedes, R.N.C. and Oliveira, A.C. 1998a. Influence of strips of native vegetation on Lepidoptera associated with *Eucalyptus cloeziana* in Brazil. Forest Ecology and Management 108, 85-90.

Zanuncio, J.C., Batalha, V.C., Guedes, R.N.C. and Picanço, M.C. 1998b. Insecticide selectivity to *Supputius cincticeps* (Stal) (Het.: Pentatomidae) and its prey *Spodoptera frugiperda* (J.E. Smith) (Lep.: Noctuidae). Journal of Applied Entomology 122, 457-460.

Zanuncio, J.C., Guedes, R.N.C., Zanuncio, T.V. and Fabres, A.S. 2001a. Species richness and abundance of defoliating Lepidoptera associated with *Eucalyptus grandis* in Brazil and their response to plant age. Austral Ecology 26, 582-589.

Zanuncio, J.C., Molina-Rugama, A.J., Serrao, J.E. and Pratissoli, D. 2001b. Nymphal development and reproduction of *Podisus nigrispinus* (Heteroptera: Pentatomidae) fed with combinations of *Tenebrio molitor* (Coleoptera: Tenebrionidae) pupae and *Musca domestica* (Diptera: Muscidae) larvae. Biocontrol Science and Technology 11, 331-337.

Zanuncio, J.C., Pereira, F.F., Jacques, G.C., Tavares, M.T. and Serrao, J.E. 2008. *Tenebrio molitor* Linnaeus (Coleoptera: Tenebrionidae), a new alternative host to rear the pupae parasitoid *Palmistichus elaeisis* Delvare & Lasalle (Hymenoptera: Eulophidae). The Coleopterists Bulletin 62, 64-66.

Zanuncio, J.C, Beserra, E.B., Molina-Rugama, A.J., Zanuncio, T.V., Pinon, T.B.M. and Maffia, V.P. 2005. Reproduction and longevity of *Supputius cincticeps* (Het.: Pentatomidae) fed with larvae of *Zophobas confusa*, *Tenebrio molitor* (Col.: Tenebrionidae) or *Musca domestica* (Dip.: Muscidae). Brazilian Archives of Biology and Technology 48, 771-777.

CHAPTER I

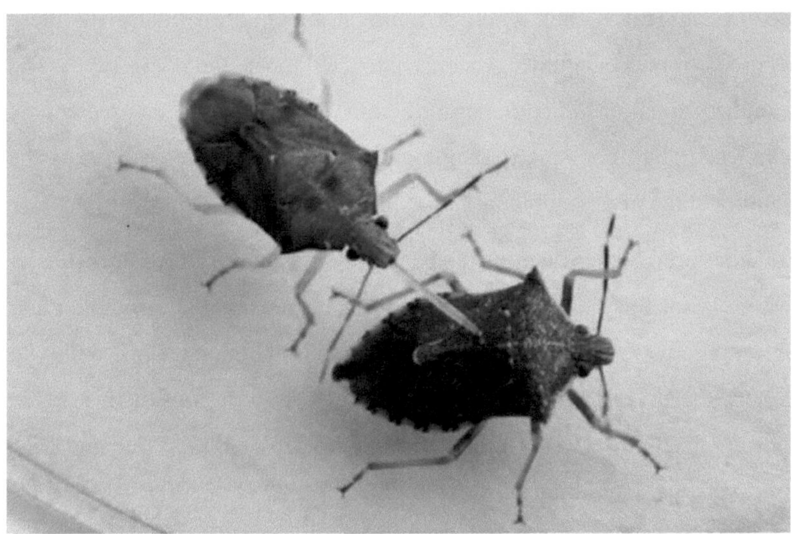

Predation and food consumption in intra- and interspecific interactions between adults of *Podisus nigrispinus* and *Supputius cincticeps* (Heteroptera: Pentatomidae)

INTRODUCTION

Predatory heteropterans are commonly found in various crops feeding on insect larvae and have a high potential for biological control (Medeiros *et al.*, 2003). *Podisus nigrispinus* (Dallas) and *Supputius cincticeps* (Stâl) (Heteroptera: Pentatomidae) are generalist predators from the Neotropical region of South America and consume a wide range of prey, mainly insects from the orders Lepidoptera, Coleoptera, Diptera and Hemiptera (Lemos *et al.,* 2003). These natural enemies are generalists and share common prey, and may prey on herbivores, saprophages or carnivores (Lemos *et al.,* 2003).

Predators locate herbivores by vibrations produced by the pest during feeding, by fecal residues and odors released by the pest or chemical clues from damaged plants (Pfannenstiel *et al.*, 1995; Sabelis *et al.*, 2001). The foraging behavior of *Podisus maculiventris* (Say) (Heteroptera: Pentatomidae) is guided by vibrations produced by prey when feeding on leaves, but when arriving at the microhabitat with these herbivores, this predator may encounter other predators (Pfannenstiel *et al.*, 1995). The occurrence of unidirectional or bidirectional intraguild predation between predators is common and can lead to harm or benefit to biological control, being determined by the mobility, size and/or stage of development of the predators (Lucas, 2005).

Interaction with other predators influences the behavioral response of predators (Lester and Harmsen, 2002). For this reason, identifying the mechanisms that promote these synergistic or antagonistic interactions between natural enemies is vital to understanding how the diversity of natural enemies affects pest control in a given system (Meisner *et al.*, 2010). Two of these mechanisms are predation rate and food consumption per predator (Bjorkman and Liman, 2005; Wilby *et al.*, 2005).

Asopinae predators have an attack mechanism that consists of inserting the stylet into the body of the prey, causing progressive paralysis and allowing it to feed. Paralysis of the prey is part of the feeding mechanism known as extra-oral digestion (Azevedo *et al.,* 2007). The advantages of this digestion are expressed ecologically by the reduction in prey handling time and the increase in nutrients ingested by the predator (Cohen, 1995). This strategy is used by predators to consume relatively large prey in relation to their size (Cohen, 1995). Foraging and feeding strategies differ between natural enemies in order to balance the risks and benefits of choosing a prey according to its biology and foraging environment (Bennett *et al.*, 2009). Variations in a predator's feeding pattern may be related to the distribution of resources and competition (Amaresekare, 2003; Choh *et al.*, 2010) or to the prey's defense mechanisms (Soares *et al*., 2009ab; Nakazawa et *al*., 2010). Antagonistic interactions can appear at low prey densities, while synergism often occurs at high prey densities (Grez *et al*., 2011)

The amount of food consumed is essential for basic needs such as growth, development and reproduction (Pereira *et al.*, 2010). Consumption is related to prey preparation and predator hunger (Pereira *et al.,* 2010). Prey preparation involves attacking, injecting toxins and enzymes to digest the tissues and ingesting the liquefied contents (Azevedo *et al.,* 2007). In addition, the amount of food consumed depends on the weight of the predator and can be affected by competition (Pereira *et al.*, 2008).

The rate of predation depends on the success of the attacks, the time the prey is exposed to the predator and the time taken to manipulate it. This manipulation, i.e. the time spent capturing, ingesting and digesting the prey, affects the predator's ability to extract the food (Pereira *et al.*, 2008). These competencies reflect differences in the response of predators and their prey, and each predator can have different types of behavior in relation to prey density and interactions with other predators (Lane *et al.*, 2006). Efficient predators must adapt to drastic changes in prey density in order to survive during periods of scarcity (Molina-Rugama *et al.*, 1998). *Podisus nigrispinus* and *S'. cincticeps* have been studied, but there is little information on the feeding behavior of these predators together.

The aim of this study was to evaluate the effects of the trophic interaction between the predators *P. nigrispinus* and *S. cincticeps*, through the rate of predation and food consumption and to observe the occurrence of cannibalism and intraguild predation and their effects on the survival and longevity of these predators.

MATERIAL AND METHODS

The experiments were carried out at the Insect Biological Control Laboratory of the Institute of Biotechnology Applied to Agriculture (BIOAGRO) of the Federal University of Viçosa in Viçosa, Minas Gerais, Brazil.

Breeding *Podisus nigrispinus and Supputius cincticeps* (Heteroptera: Pentatomidae)

The predatory bedbugs were obtained from the mass rearing facility of the Insect Biological Control Laboratory of the Institute of Biotechnology Applied to Agriculture (BIOAGRO) of the Federal University of Viçosa in Viçosa, Minas Gerais, Brazil.

Adults of *P. nigrispinus* and *S. cincticeps* were placed in wooden cages (30 x 30 x 30 cm), with shingled sides and glass lids, fed on pupae of *Tenebrio molitor* L. (Coleptera: Tenebrionidae) and moistened cotton as a source of water. The eggs of these bugs were collected every two days and placed in Petri dishes (9.0 x 1.5 cm). Nymphs of these predators were reared in these Petri dishes until the fourth stage with *T. molitor* pupae and moistened cotton as a water source. After reaching the fourth stage, these nymphs were placed in wooden cages (30 x 30 x 30 cm) with shingled sides and glass lids, and *T. molitor* pupae and moistened cotton as a water source. The newly emerged adults were transferred to the cages. Rearing took place in an air-conditioned room at 25 ± 2°C, relative humidity of 60 ± 10% and 12-hour photophase (Zanuncio *et al.*, 2001).

Setting up the experiment

The experiment was carried out in the Insect Biological Control Laboratory in 500 mL plastic jars with a hole in the lid sealed with organza to increase aeration and a plastic tube with 2.5 ml of distilled water, sealed with cotton, to supply water and humidity (Lemos *et al.*, 2003). The jars were kept at 25 ± 1°C, 70 ± 10% relative humidity and 12-hour photophase in a randomized block design with two factors in a 5 x 2 scheme and fifteen replications.

One factor was the species and number of female bedbugs with five levels: Tl- one female of *P. nigrispinus*; T2- two females *of P. nigrispinus*; T3- one female of *S. cincticeps*; T4- two females of *S. cincticeps*; T5- one female of *P. nigrispinus* and another of *S. cincticeps.* The predators were left without food for 24 hours and were weighed on a precision analytical balance (±0.1 mg) before the start of the experiment (Pereira *et al.,* 2008). Newly emerged females *of S. cincticeps* weighing between 40.0 and 50.0 mg and of *P. nigrispinus* between 55.0 and 65.0 mg were selected to reduce the influence of weight.

The other factor was the type of food with two levels: Dl- no food; D2- one *T. molitor* pupa (weight 105 ± 5 mg) per predator. The Dl level was used to observe the occurrence or not of cannibalism and intraguild predation and their influence on the survival and longevity of the predators, until the predators died.

T. molitor pupae were weighed and offered to the predators every 24 hours. The weight of the bedbugs and pupae and the amount of food consumed were recorded every 24 hours for nine days (Pereira *et al.*, 2008). The pupae removed from each day were stored in Petri dishes to confirm their death or not. Consumption was calculated by subtracting the weight of the pupae before they were offered to the predators from the weight after 24 hours of exposure to the predators. Fifteen pupae per treatment were weighed and kept without predation for 24 hours. These pupae served as a control and the calculation of their biomass allowed natural weight loss to be discounted.

Statistical analysis

The predation rate, total food consumption and after 24 hours, total weight gain and after 24 hours of the predators, the occurrence of intraguild predation and cannibalism and their effects on predator longevity were evaluated. The data was submitted to analysis of variance using the F test and the means were compared using the Tukey test at a 5% significance level, when applicable, using the "R" statistical program (R Development Core Team, 2010).

RESULTS

The survival of *T. molitor* pupae was 100% in the absence of predators, so there was no need to correct the mortality data for this prey. The natural weight loss of the pupae after 24 hours was discounted when calculating food consumption.

Intraspecific interactions

Food consumption per individual of *T. molitor* pupae after 24 hours of exposure did not differ between treatments with one or two females of *P. nigrispinus* (F= 0.2585; P= 0.6156). There was also no difference between food consumption at the end of the nine days in these treatments (F= 1.0232; P= 0.3223). The predation rate was also similar between treatments (F= 0.2314; P= 0.6347) (Table 1).

Table 1 - Means ± standard error[1] of food consumption, per individual, after 24 hours and total over nine days (mg) and total predation rate (%) of female adults of *Podisus nigrispinus* (Heteroptera: Pentatomidae) feeding on pupae of *Tenebrio molitor* (Coleoptera: Tenebrionidae) at 25 ± 1°C, 70 ± 10% relative humidity and 12-hour photophase. Treatments: 2PN X 2TM= two females *of P. nigrispinus* X two pupae of *T. molitor*; PN X TM= one female of *P. nigrispinus* X one pupa of *T. molitor*.

Treatment	Consumption (24h)	Total consumption	Predation rate (%)
2PN X 2TM	39,06 ± 6,19a	112,03 ± 9,65a	76,04 ± 4,54a
PN X TM	42,50 ± 3,49a	101,31 ± 5,82a	72,50 ± 5,48a

(1) Averages followed by the same small letter per column do not differ by the "F" test at 5% probability.

Food consumption per individual of *T. molitor* pupae after 24 hours of exposure differed between treatments with one or two females of *S cincticeps*. Consumption was lower with two females of this species (F= 10.148; P= 3.532 x 10^{-3}), but at the end of nine days it did not differ between these treatments (F= 0.4847; P= 0.4922). The predation rate was also similar between treatments (F= 0.009; P= 0.925) (Table 2).

Table 2 - Means ± standard error[1] of food consumption, per individual, after 24 hours and total over nine days (mg), and total predation rate (%) of female adults of *Supputius cincticeps* (Heteroptera: Pentatomidae) feeding on pupae of *Tenebrio molitor* (Coleoptera: Tenebrionidae) at 25 ± 1°C, 70 ± 10% relative humidity and 12-hour photophase. Treatments: 2SC X 2TM= two *S. cincticeps* females X two *T. molitor* pupae; SC X TM= one *S. cincticeps* female X one *T. molitor* pupa.

Treatment	Consumption (24h)	Total consumption	Predation rate (%)
2SC X 2TM	14,45 ± 2,42b	48,99 ± 4,62a	50,00 ± 3,61a
SC X TM	24,83 ± 2,37a	52,83 ± 2,84a	49,52 ± 3,48a

(1) Averages followed by the same small letter per column do not differ by the "F" test at 5% probability.

Interspecific interactions

Food consumption per individual of *T. molitor* pupae after 24 hours of exposure did not differ between treatments with one female of *P. nigrispinus* and one of *S. cincticeps* preying together or separately (F= 0.906; P= 0.7645). Food consumption at the end of the nine days was also similar in these treatments (F= 0.898; P= 0.3481). The predation rate was higher with one female of *P. nigrispinus* and another of *S. cincticeps* preying together (F= 9.7175; P= 3.272 X 10^{-5}) (Table 3).

Table 3 - Means ± standard error[1] of food consumption, per individual, after 24 hours and total over nine days (mg), and total predation rate (%) of female adults of *Podisus nigrispinus* and *Supputius cincticeps* (Heteroptera: Pentatomidae) feeding on pupae of *Tenebrio molitor* (Coleoptera: Tenebrionidae) at 25 ± 1°C, 70 ± 10% relative humidity and 12-hour photophase. Treatments: PN + SC X 2TM= one female of *P. nigrispinus* + one female of *S. cincticeps* X two pupae of *T. molitor*; (PN X TM) + (SC X TM)= one female of *P. nigrispinus* X one pupa of *T. molitor* + one female of *S. cincticeps* X one pupa of *T. molitor*.

Treatment	Consumption (24h)	Total consumption	Predation rate (%)
PN + SC X 2TM	32,37 ± 2,57a	83,43 ± 4,83a	72,75 ± 4,02a
(PN X TM) + (SC X TM)	33,75 ± 2,11a	76,47 ± 2,27a	61,03 ± 3,07b

(1) Averages followed by the same small letter per column do not differ by the "F" test at 5% probability.

The predation rate of one female of *P. nigrispinus* and one female of *S'. cincticeps* preying together was similar to that of two females *of P. nigrispinus* (F= 0.3537; P= 0.5545), and higher than that of two females *of S. cincticeps* preying together (F= 9.7175; P= 3.27 X 10^{-5}) (Figure 1).

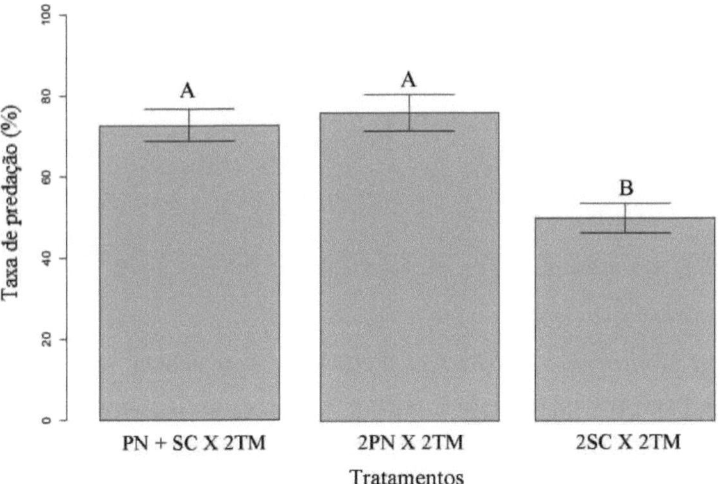

Figure 1 - Predation rate (%) by females *of Podisus nigrispinus* and *Supputius cincticeps* (Heteroptera: Pentatomidae) on pupae *of Tenebio molitor* (Coleoptera: Tenebrionidae) at 25 ± 1°C, 70 ± 10% relative humidity and 12-hour photophase. Treatments: PN + SC X 2TM= one female of

P. nigrispinus + one female of *S. cincticeps* X two pupae of *T. molitor*; 2PN X 2TM= two females of *P. nigrispinus* X two pupae of *T. molitor*; 2SC X 2TM= two females of *S. cincticeps* X two pupae of *T. molitor*.

Weight gain

The weight gain per *P. nigrispinus* female, after exposure for 24 hours to *T. molitor* pupae, did not differ between treatments with one *P. nigrispinus* female, two *P. nigrispinus* females or one P. nigrispinus female and one *S. cincticeps* female preying together (F= 0.6647; P= 0.5188). There was also no difference in weight gain at the end of nine days in these treatments (F= 0.7171; P= 0.4937) (Table 4). Spawning *of P. nigrispinus* was observed in all treatments

Table 4 - Means ± standard error[1] of weight gain after 24 hours and total over nine days (mg) of female adults of *Podisus nigrispinus* (Heteroptera: Pentatomidae) feeding on pupae *of Tenebrio molitor* (Coleoptera: Tenebrionidae) at 25 ± 1°C, 70 ± 10% relative humidity and 12-hour photophase. Treatments: PN X TM= one female of *P. nigrispinus* X one pupa of *T. molitor*; 2PN X 2TM= two females *of P. nigrispinus* X two pupae of *T. molitor*; PN + SC X 2TM= one female of *P. nigrispinus* + one female of *Supputius cincticeps* (Heteroptera: Pentatomidae) X two pupae of *T. molitor*.

Treatment	Weight gain (24h)	Total weight gain
PN X TM	22,62 ± 1,85a	22,89 ± 2,66a
2PN X 2TM	20,47 ± 2,91a	27,54 ± 2,08a
PN + SC X 2TM	24,79 ± 2,52a	27,92 ± 3,93a

(1) Averages followed by the same small letter per column do not differ by the "F" test at 5% probability.

The weight gain per *S. cincticeps* female was highest in the treatment with one *S. cincticeps* female and lowest with two (F= 4.4577; P= 0.03914). However, the weight gain per *S. cincticeps* female at the end of nine days was similar between these treatments (Table 5).

Table 5 - Means ± standard error[1] of weight gain after 24 hours and total over nine days (mg) of female adults of *Supputius cincticeps* (Heteroptera: Pentatomidae) feeding on pupae of *Tenebrio molitor* (Coleoptera: Tenebrionidae) at 25 ± 1°C, 70 ± 10% relative humidity and 12-hour photophase. Treatments: SC X TM= one female of *S. cincticeps* X one pupa of *T. molitor*; 2SC X 2TM= two females *of S. cincticeps* X two pupae of *T. molitor*; SC + PN X 2TM= one female of *S. cincticeps* + one female of *Podisus cincticeps* (Heteroptera: Pentatomidae) X two pupae of *T. molitor*.

Treatment	Weight gain (24h)	Total weight gain
SC X TM	12,68 ± 1,96a	9,99 ± 2,14a
2SC X 2TM	7,12 ± 1,44b	7,54 ± 1,61a
SC + PN X 2TM	9.97 ± 2.54ab	5,48 ± 4,47a

(1) Averages followed by the same small letter per column do not differ by the "F" test at 5% probability.

Cannibalism and intraguild predation

Cannibalism and intraguild predation were observed only in the treatments without food. Intraguild predation was bidirectional, i.e. females *of P. nigrispinus* preyed on females *of S. cincticeps* and vice versa (Figure 2). However, cannibalism and intraguild predation were higher in *S. cincticeps* females (Figures 3 and 4).

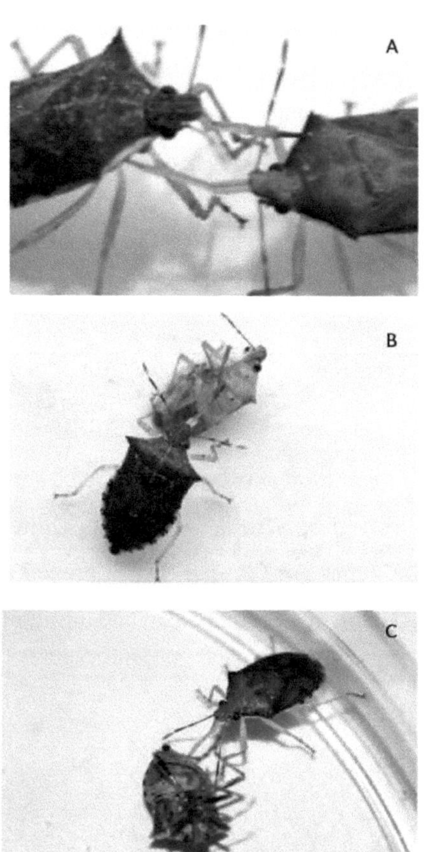

Figura 2 - Intraguild predation between females of *Podisus nigrispinus and Supputius cincticeps* (Heteroptera: Pentatomidae). A- Bidirectional intraguild predation between *P. nigrispinus* and *S. cincticeps*; B- Intraguild predation of *P. nigrispinus* on *S. cincticeps*; C- Intraguild predation of *S. cincticeps* on *P. nigrispinus*.

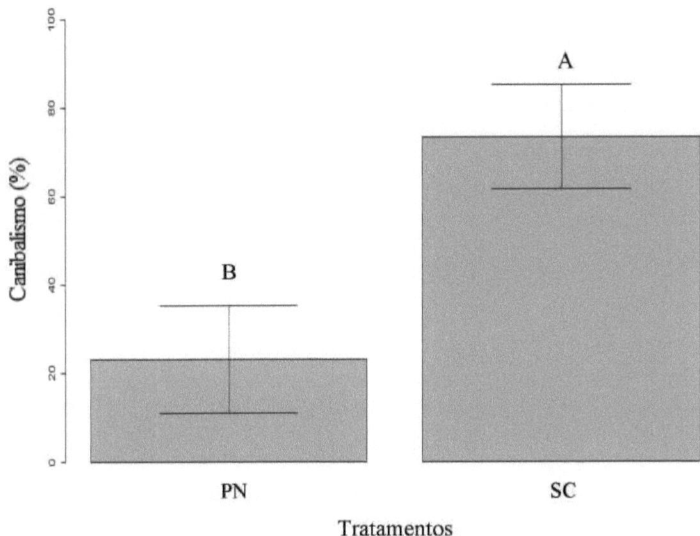

Figura 3 - Cannibalism (%) between females of *Podisus nigrispinus* (PN) and *Supputius cincticeps* (SC) (Heteroptera: Pentatomidae) without feeding at 25 ± 1°C, 70 ± 10% relative humidity and 12-hour photophase.

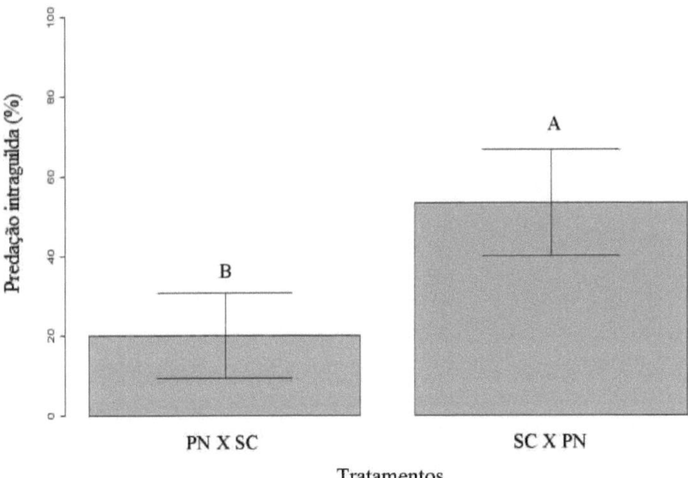

Figura 4 - Intraguild predation (%) of *Podisus nigrispinus* (Heteroptera: Pentatomidae) females on *Supputius cincticeps* (PN X SC) (Heteroptera: Pentatomidae) females and *S. cincticeps* females on *P. nigrispinus* (SC X PN) females without feeding at 25 ± 1°C, 70 ± 10% relative humidity and 12-hour photophase.

The longevity of *P. nigrispinus* females did not differ between treatments without food, with cannibalism and intraguild predation (F= 0.4066; P= 0.6719) (Figure 5). However, the longevity of *S. cincticeps* females with cannibalism or intraguild predation was greater than without food (F= 3.4023; P= 0.04615) (Figure 6).

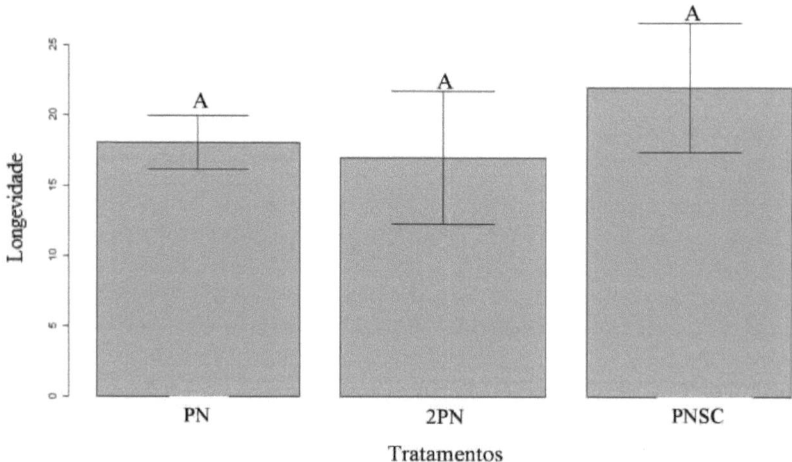

Figura 5 - Longevity (days) of *Podisus nigrispinus* (Heteroptera: Pentatomidae) females with cannibalism and intraguild predation at 25 ± 1°C, 70 ± 10% relative humidity and 12-hour photophase. Treatments: PN= no feeding; 2PN= cannibalism; PNSC= intraguild predation.

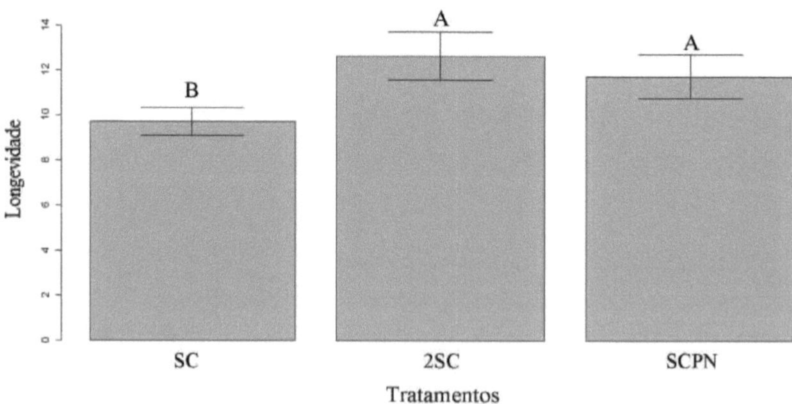

Figura 6 Longevity (days) of females *of Supputius cincticeps* (Heteroptera: Pentatomidae) with cannibalism and intraguild predation at 25 ± 1°C, 70 ± 10% relative humidity and 12-hour

photophase. Treatments: SC= no feeding; 2SC= cannibalism; SCPN= intraguild predation.

DISCUSSION

Intraspecific interactions

Intraspecific interactions between females *of P. nigrispinus* were not enough to affect the food consumption and predation rate of this predator. This is because the high availability of food reduces cannibalism and competition (Sato *et al.*, 2009; Ware *et al.*, 2009). One pupa of *T. molitor* (105 ± 5 mg) satiated one female of this predator, as observed by its feeding on five caterpillars *of Alabama argillacea* (Hübner) (Lepidoptera: Noctuidae) (71.37 mg) (Pereira *et al.,* 2008).

Intraspecific interactions between *S. cincticeps* females should not have affected this predator's food consumption and predation rate, as one *T. molitor* pupa can satiate this predator, and the amount of food consumed depends on the predator's weight (Pereira *et al.*, 2008) and *P. nigrispinus* females are larger than *S. cincticeps* females. Intraspecific competition affected food consumption after 24 hours, when the predator was hungry (24 hours without food). Interspecific competition also affected the food consumption of *Stegodyphus dumicola* Pocock (Araneae: Eresidae) (Amir *et al.*, 2000). In addition to competition, a high rate of cannibalism among adults of this species can lead to anti-predation behavior, reducing their food consumption (Werner and Anholt, 1996). This reduction is a common response, as it reduces the risk of predation (Relyea, 2004; Jara and Perotti, 2010). Cannibalism was only observed in situations of hunger and did not occur throughout the experiment, so food consumption at the end of the nine days and the rate of predation were not affected.

Interspecific interactions

Interspecific interactions between females *of P. nigrispinus* and *S. cincticeps* have not been sufficient to affect the food consumption of these predators, due to the high availability of food, which decreases intraguild predation and competition (Ware *et al.*, 2009; Chacón and Heimpel, 2010). This high food availability often leads to synergistic effects on predation rates (Grez *et al.*, 2011), as observed between *P.*

nigrispinus and *S. cincticeps*. *Orthotylus marginalis* Reuter (Heteroptera: Miridae) and *Anthocoris nemorum* (L.) (Heteroptera: Anthocoridae) preying on *Phratora vulgatissima* L. (Coleoptera: Chrysomelidae) showed a positive response in the interspecific interaction, explained, in part, by a preference for different stages of prey development (Bjorkman and Liman, 2005). The effect of the interaction between *Podisus maculiventris* (Say) (Heteroptera: Pentatomidae) and *Coleomegilla maculata* Lengi (Coleoptera: Coccinellidae) preying *on Leptinotarsa decemlineata* Say (Coleoptera: Chrysomelidae) was antagonistic, mainly due to intraguild predation (Mallampalli *et al.*, 2002). The predation rate increased as *P. nigrispinus*, with *S. cincticeps*, preyed on more than one *T. molitor* pupa per day. Even when satiated, this predator can attack prey with high abundance (Vivian *et al.*, 2002), as shown for *P. maculiventris* and *Podisus scaglila* (Fabricius) (Heteroptera: Pentatomidae), which preyed on more prey than necessary to satisfy their hunger (De Clercq and Degheele, 1994). In addition, predation behavior can make the prey more susceptible to the other predator, a phenomenon called predator facilitation (Charnov *et al.*, 1976; Losey and Denno, 1998).

The similar predation rate of *P. nlgrlsplnus* plus *S. clnctlceps* and two females of the first predator shows that the use of just one species can be more effective in biological control than a combination of predators (Straub and Snyder, 2006).

The weight of *T. molltor* pupae shows that each one can satiate a female of *P. nlgrlsplnus* or *S. clnctlceps* and it was impossible to carry out the experiment at low prey densities. Antagonistic interactions appear under these conditions, while synergism often occurs at high prey densities (Grez *et al.*, 2011). In addition, the experiment was short-term, and the mechanisms of negative diversity effects, such as intraguild predation, occur at the population level and generally involve changes in prey abundance, which can take some time to manifest (Rosenheim, 2001).

Weight gain

The greater weight gain of *P. nigrispinus*, after 24 hours and at the end of nine days, in all treatments, is important due to the body mass greater than 60 mg, which is the weight range for adequate fecundity of this predator (Zanuncio *et al.* 2002). Females showed a similar weight gain after 24 hours and at the end of nine days, indicating that there was no effect of interspecific or intraspecific competition. Similar weight gain after 24 hours was observed for *P. nigrispinus* preying *on Alabama argillacea* (Hübner) (Lepidoptera: Noctuidae) (Oliveira *et al.*, 2001). However, in another experiment, the weight gain of *P. nigrispinus* preying *on A. argillacea* was lower after 24 hours (Pereira *et al.,* 2008). This is due to the use of predator females and lower weight prey, since smaller prey consume less food and because they prey on smaller prey, they need to prey on more prey, spending more energy searching for them. The presence of eggs in all treatments shows that the females were satiated, as satiated predators of the *Podisus* genus can produce infertile eggs, even when not mated (De Clercq and Degheele, 1990).

The similar weight gain between treatments with *S. cincticeps* at the end of nine days shows that there was no effect of interspecific or intraspecific competition. Intraspecific competition, after 24 hours, reduced the weight gain of this predator, due to a decrease in consumption. The risk of being preyed upon, due to the high rate of cannibalism among adults of this species, may lead one predator to avoid the other, expending more energy. This behavior is common and reduces the risk of predation (Relyea, 2004; Jara and Perotti, 2010). Cannibalism was only observed in starvation situations, which is why food consumption at the end of the nine days was not affected.

Cannibalism and intra-guild predation

Cannibalism and intraguild predation in conditions of lack of food is due to the fact that predatory bedbugs can select prey with less defense potential in conditions of abundance, reducing unnecessary energy expenditure (Zanuncio *et al*., 2008). Adults of *P. maculiventris* prey on adults of *Coccinella septempunctata* Linnaeus (Coleoptera: Coccinellidae) only in the absence of food (Moreno *et al*., 2010). Bidirectional intraguild predation occurred, but intraguild predation by females *of S. cincticeps* was greater, even though this species is smaller than *P. nigrispinus*. The relative size of predators affects the outcome of predation (Felix and Soares, 2004; Straub *et al.*, 2008) and larger species generally prey on smaller ones. The higher rate of predation and cannibalism by unfed females *of S, cincticeps* is due to their shorter lifespan than those of *P, nigrispinus*, requiring them to feed earlier in order to survive. *H. axyridis* larvae were more aggressive as hunger increased (Burgio *et al*., 2002; Moser and Obrycki, 2009; Moser *et al*., 2010).

Consumption of *S. cincticeps* showed no energy gain for *P. nigrispinus*, because, despite consuming nutrients, the energy cost of capturing and killing intraguild prey is expensive, as it can counterattack and the same occurs for cannibalism. This explains the low rate of these forms of feeding for *P, nigrispinus*. Larger larvae *of Harmonia axyridis* (Pallas) (Coleoptera: Coccinellidae) counter-attack *P, maculiventris* as a form of defense (De Clercq *et al*., 2003). Intraguild predation and cannibalism increased the longevity *of S. cincticeps*. These forms of feeding increase the ability of predators to persist in an environment with low food resources (Ruberson *et al*., 1986; Rudolf and Armstrong, 2008; Culler and Lamp, 2009). In the field, rates of cannibalism and intraguild predation are unlikely to reach levels similar to those in the laboratory, due to the possibility of insect dispersal and the presence of alternative prey. More complex habitats generally contain a greater diversity of food resources and refuges for predators, reducing the intensity of competition, cannibalism and intraguild predation (Abad-Moyano *et al*., 2010; Frank *et al,* 2010; Korenko and Pekàr, 2010).

Predators can coexist in the same environment when the intraguild prey is more efficient in competing for the shared resource, while the intraguild prey is more efficient in competing for the shared resource.

The intraguild predator compensates for this lack of competitiveness by consuming the intraguild prey (Holt and Polis, 1997; Borer *et al*., 2003), so we can conclude that *P. nigrispinus* and *S. cincticeps* can coexist in the same environment, since *P. nigrispinus* is the intraguild prey and best competitor and *S. cincticeps* is the intraguild predator. This can reduce predation of the target prey, due to intraguild predation (Polis *et al*., 1989) and anti-predation behavior, which can lead the intraguild prey to avoid areas with the predator (Magalhaes *et al*., 2004; Çakmak *et al*., 2006). Increasing the density of the target prey depends on the degree to which the intraguild predator feeds on it, which can reduce top-down effects on herbivores, weaken trophic cascades, and impair biological control (Rosenheim *et al.*, 1995; Denno *et al.*, 2004).

CONCLUSION

Intra- and interspecific interactions did not affect *P. nigrispinus* females with a high prey density. Even in the absence of food, they are little affected by these interactions, as they have low cannibalism, but can be preyed upon by females *of S. cincticeps*.

Females *of S. cincticeps*, with a high prey density, were also not affected by intra- and interspecific interactions. In the absence of food, they benefit from these interactions, as they are little preyed upon by *P. nigrispinus* and have a high occurrence of cannibalism and intraguild predation, which increases their longevity.

At high prey density, there was synergism in the predation rate between these predators, which is beneficial for biological control. The coexistence *of P. nigrisipinus* and *S'. cincticeps* is possible in the same environment.

The conservation of both predators is interesting in conservative biological control programs, as their coexistence is possible and they are synergistic at high prey densities. However, it is better to use only *P. nigrispinus* in inundative biological control programs, as this predator has a higher predation rate, a low cannibalism rate and greater longevity when prey is scarce. In addition, the combined use with *S. cincticeps*, an intraguild predator, could increase the mortality of *P. nigrispinus* or cause it to migrate, jeopardizing biological control. For this reason, it is better to invest financial resources in producing a greater number of *P. nigrispinus* individuals.

REFERENCES

Abad-Moyano, R., Urbaneja, A. and Schausberger, P. 2010. Intraguild interactions between *Euseius stipulates* and the candidate biocontrol agents of *Tetranychus urticae* in Spanish clementine orchards: *Phytoseiulus persimilis* and *Neoseiulus californicus*. Experimental & Applied Acarology 50, 23-34.

Amarasekare, P. 2003. Competitive coexistence in spatially structured environments: a synthesis. Ecology Letters 6, 1109-1122.

Amir, N., Whitehouse, M.E.A. and Lubin, Y. 2000. Food consumption rates and competition in a communally feeding social spider, *Stegodyphus dumicola* (Eresidae). Journal of Arachnology 28, 195-200.

Azevedo, D.O., Zanuncio, J.C., ZanuncioJunior, J.S., Martins, G.F., Marques- Silva, S., Sossai, M.F. and Serrao, J.E. 2007. Biochemical and morphological aspects of the predator *Brontocoris tabidus* (Heteroptera: Pentatomidae). Brazilian Archives of Biology and Technology 50, 469-477.

Bennett, J.A., Gillespie, D.R., Shipp, J.L. and Vanlaerhoven, S.L. 2009. Foraging strategies and patch distributions: intraguild interactions between *Dicyphus hesperus* and *Encarsia formosa*. Ecological Entomology 34, 58-65.

Bjorkman, C. and Liman, A.S. 2005. Foraging behavior influences the outcome of predator-predator interactions. Ecological Entomology 30, 164-169.

Borer, E.T., Briggs, C.J., Murdoch, W.W. and Swarbrick, S.L. 2003. Testing intraguild predation theory in a field system: does numerical dominance shift along a gradient of productivity? Ecology Letters 6, 929-935.

Burgio, G., Santi, F. and Maini, S. 2002. On intra-guild predation and cannibalism in *Harmonia axyridis* (Pallas) and *Adalia bipunctata* L. (Coleoptera: Coccinellidae). Biological Control 24, 110-116.

Çakmak, I., Janssen, A. and Sabelis, M.W. 2006. Intraguild interactions between the predatory mites *Neoseiulus californicus* and *Phytoseiulus persimilis*. Experimental and Applied Acarology 38, 33-46.

Casula, P., Wilby A. and Thomas M.B. 2006. Understanding biodiversity effects on prey in multi-enemy systems. Ecology Letters 9, 995-1004.

Chacón, J.M. and Heimpel, G.E. 2010. Density-dependent intraguild predation of an aphid parasitoid. Oecologia 164, 213-220.

Charnov, E.L., Orians, G.H. and Hyatt, K. 1976. Ecological implications of resource depression. American Naturalist 110, 247-259.

Choh, Y., van der Hammen, T., Sabelis, M.W. and Janssen, A. 2010. Cues of intraguild predators affect the distribution of intraguild prey. Oecologia 163, 335-340.

Cohen, A.C. 1995. Extra-oral digestion in predaceous terrestrial Arthropoda. Annual Review of Entomology 40, 85-103.

Culler, L.E. and Lamp, W.O. 2009. Selective predation by larval *Agabus* (Coleoptera: Dytiscidae) on mosquitoes: support for conservation-based mosquito suppression in constructed wetlands. Freshwater Biology 54, 2003-2014.

De Clercq, P. and Degheele, D. 1990. Description and life history of the predatory bug *Podisus sagitta* (Heteroptera: Pentatomidae). Canadian Entomologist 122, 1149-1156.

De Clercq, P. and Degheele, D. 1994. Laboratory measurement of predation by *Podisus maculiventris* and *Podisus sagitta* (Heteroptera: Pentatomidae) on beet armyworm (Lepidoptera: Noctuidae). Journal of Economic Entomology 87, 76-83.

Denno, R.F., Mitter, M.S., Langellotto, G.A., Gratton, C. and Finke, D.L. 2004. Interactions between a hunting spider and a web-builder: consequences of intraguild predation and cannibalism for prey suppression. Ecological Entomology 29, 566-577.

Felix, S. and Soares, A.O. 2004. Intraguild predation between the aphidophagous ladybird beetles *Harmonia axyridis* and *Coccinella undecimpunctata* (Coleoptera: Coccinellidae): the role of body weight. European Journal of Entomology 101, 237-242.

Frank, S.D., Shrewsbury, P.M. and Denno, R.F. 2010. Effects of alternative food on cannibalism and herbivore suppression by carabid larvae. Ecological Entomology 35, 61-68.

Grez, A.A., Zaviezo, T. and Mancilla, A. 2011. Effect of prey density on intraguild interactions among foliar- and ground-foraging predators of aphids associated with alfalfa crops in Chile: a laboratory assessment. Entomologia Experimentalis et Applicata 139, 1-7.

Hol, R.D. and Polis, G.A. 1997. A theoretical framework for intraguild predation. The American Naturalist 149, 745-764.

Janssen, A., Sabelis, M.W., Magalhaes, S., Montserrat, M. and van der Hammen, T. 2007. Habitat structure affects intraguild predation. Ecology 88, 27132719.

Jara, F.G. and Perotti, M.G. 2010. Risk of predation and behavioral response in three anuran species: influence of tadpole size and predator type. Hydrobiologia 644, 313-324.

Lane, S.D., St Mary, C.M. and Getz, W.M. 2006. Coexistence of attack-limited parasitoids sequentially exploiting the same resource and its implications for biological control. Annales Zoologici Fennici 43, 17-34.

Lemos, W.P., Ramalho, F.S., Serrao, J.E. and Zanuncio, J.C. 2003. Effects of diet on development of *Podisus nigrispinus* (Dallas) (Het., Pentatomidae), a predator of the cotton leafworm. Journal of Applied Entomology 127, 389-395.

Lester, P.J. and Harmsen, R. 2002. Functional and numerical responses do not always indicate the most effective predator for biological control: an analysis of two predators in a two-prey system. Journal of Applied Ecology 39, 455-468.

Losey, J.E. and Denno, R.F. 1998. Positive predator-predator interactions: enhanced predation rates and synergistic suppression of aphid populations. Ecology 79, 2143-2152.

Lucas, E. 2005. Intraguild predation among aphidophagous predators. European Journal of Entomology 102, 351-364.

Magalhaes, S., Tudorache, C., Montserrat, M., van Maanen, R., Sabelis, M.W. and Janssen, A. 2004. Diet of intraguild predators affects antipredator behavior in intraguild prey. Behavioral Ecology 16, 364-370.

Mallampalli, N., Castellanos, I. and Barbosa, P. 2002. Evidence for intraguild predation by *Podisus maculiventris* on a ladybeetle, *Coleomegilla maculata:* implications for biological control of Colorado potato beetle, *Leptinotarsa decemlineata.* BioControl 47, 387-398.

Medeiros, R.S., Ramalho, F.S., Zanuncio, J.C. and Serrao, J.E. 2003. Effect of temperature on life table parameters of *Podisus nigrispinus* (Heteroptera: Pentatomidae) fed with *Alabama argillacea* (Lepidoptera: Noctuidae) larvae. Journal of Applied Entomology 127, 209-213.

Meisner, M., Harmon, J.P., Harvey, C.T. and Ives, A.R. 2010. Intraguild predation on the parasitoid *Aphidius ervi* by the generalist predator *Harmonia axyridis*: the threat and its avoidance. Entomologia Experimentalis et Applicata 138,193-201.

Molina-Rugama, A.J., Zanuncio, J.C., Zanuncio, T.V. and De Oliveira, M.L.R. 1998. Reproductive strategy of *Podiscus rostralis* (Stal) (Heteroptera: Pentatomidae) females under different feeding intervals. Biocontrol Science and Technology 8, 583588.

Moreno, C.R., Lewins, S.A. and Barbosa, P. 2010. Influence of relative abundance and taxonomic identity on the effectiveness of generalist predators as biological control agents. Biological Control 52, 96-103.

Moser, S.E. and Obrycki, J.J. 2009. Competition and intraguild predation among three species of coccinellids (Coleoptera: Coccinellidae). Annals of the Entomological Society of America 102, 419-425.

Moser, S.E., Haynes, K.F. and Obrycki, J.J. 2010. Behavioral response to larval tracks and the influence of tracks on intraguild scavenging by coccinellid larvae. Journal of Insect Behavior 23, 45-58.

Nakazawa, T., Miki, T. and Namba, T. 2010. Influence of predator-specific defense adaptation on intraguild predation. Oikos 119, 418-427.

Oliveira, J.EM., Torres, J.B., Moreira, A.F.C. and Zanuncio, J.C. 2001. Effect of prey density and mating status on predation rate of females *of Podisus nigrispinus* (Dallas) (Heteroptera: Pentatomidae) in laboratory and field conditions. Neotropical Entomology 30, 647-654.

Korenko, S. and Pekàr, S. 2010. Is there intraguild predation between winteractive spiders (Araneae) on apple tree bark? Biological Control 54, 206-212.

Pereira, A.I.A., Ramalho, F.S., Malaquias, J.B., Bandeira, C.M., Silva, J.P.S. and

Zanuncio, J.C. 2008. Density of *Alabama argillacea* larvae affects food extraction by females of *Podisus nigrispinus*. Phytoparasitica 36, 84-94.

Pereira, A.I.A., Ramalho, F.S., Rodrigues, K.C.V., Malaquias, J.B., Souza, J.V.S and Zanuncio, J.C. 2010. Food extraction by the males of *Podisus nigrispinus* (Dallas) (Hemiptera: Pentatomidae) from cotton leafworm larvae. Brazilian Archives of Biology and Technology 53, 1027-1035.

Pfannenstiel, R.S., Hunt, R.E. and Yeargan, K.V. 1995. Orientation of a hemipteran predator by feeding caterpillars. Journal of Insect Behavior 8, 1-9.

Polis, G.A., Myers, C.A. and Holt, R.D. 1989. The ecology and evolution of intraguild predation: potential competitors that eat each other. Annual Review of Entomology 20, 297-330.

R Development Core Team. 2010. R: A language and environment for statistical computing. R Foundation for Statistical Computing, Vienna, Austria. ISBN 3-900051-07-0, URL http://www.R-project.org.

Relyea, R. A. 2004. Fine-tuned phenotypes: tadpole plasticity under 16 combinations of predators and competitors. Ecology 85, 172-179.

Rosenheim, J. A. 2001. Source-sink dynamics for a generalist insect predator in habitats with strong higher-order predation. Ecological Monographs 71, 93-116.

Rosenheim, J.A., Kaya, H.K., Ehler, L.E., Marois, J.J. and Jaffee, B.A. 1995. Intraguild predation among biological control agents: theory and evidence. Biological Control 5, 303-335.

Ruberson, J.R., Tauber, M.J. and Tauber, C.A. 1986. Plant feeding by *Podisus maculiventris* (Heteroptera: Pentatomidae): effect on survival, development, and preoviposition period. Environmental Entomology 15, 894-897

Rudolf, V.H.W. and Armstrong, J. 2008. Emergent impacts of cannibalism and size refuges in prey on intraguild predation systems. Oecologia 157, 675-686.

Sabelis, M.W., Janssen, A. and Kant, M.R. 2001. The enemy of my enemy is my ally.

Science 291, 2104-2105.

Sato, S., Shinya, K., Yasuda, H., Kindlmann, P. and Dixon, A.F.G. 2009. Effects of intra and interspecific interactions on the survival of two predatory ladybirds (Coleoptera: Coccinellidae) in relation to prey abundance. Applied Entomology and Zoology 44, 215-221.

Soares, M.A., Zanuncio, J.C., Leite, G.L.D., Wermelinger, E.D. and Serrao, J.E. 2009a. Does *Thyrinteina arnobia* (Lepidoptera: Geometridae) use different defense behaviors against predators? Journal of Plant Diseases and Protection 116, 30-33.

Soares, M.A., Torres-Gutierrez, C., Zanuncio, J.C., Pedrosa, A.R.P. and Lorenzon, A.S. 2009b. Superparasitism *of Palmistichus elaeisis* (Hymenoptera: Eulophidae) and defense behavior of two hosts. Revista Colombiana de Entomologia 35, 62-65.

Straub, C.S., Finke, D.L. and Snyder, W.E. 2008. Are the conservation of natural enemy biodiversity and biological control compatible goals? Biological Control 45, 225-237.

Straub, C.S. and Snyder, W.E. 2006. Species identity dominates the relationship between predator biodiversity and herbivore suppression. Ecology 87, 277-282.

Vivian, L.M., Torres, J.B., Veiga, A.F.S.L. and Zanuncio, J.C. 2002. Predation behavior and food conversion of *Podisus nigrispinus* on the tomato moth. Pesquisa Agropecuària Brasileira 37, 581-587.

Ware, E.R., Yguel, B. and Majerus, M. 2009. Effects of competition, cannibalism and intra-guild predation on larval development of the European coccinellid *Adalia bipunctata* and the invasive species *Harmonia axyridis.* Ecological Entomology 34, 12-19.

Werner, E.E. and Anholt, B.R. 1996. Predator-induced behavioral indirect effects: consequences to competitive interactions in anuran larvae. Ecology 77, 157169.

Wilby, A., Villareal, S.C., Lan, L.P., Heong, K.L. and Thomas, M.B. 2005. Functional benefits of predator species diversity depend on prey identity. Ecological Entomology 30, 497-501.

Zanuncio, J.C., Molina-Rugama, A.J., Serrlo, J.E. and Pratissoli, D. 2001. Nymphal development and reproduction of *Podisus nigrispinus* (Heteroptera: Pentatomidae) fed with combinations of *Tenebrio molitor* (Coleoptera: Teneobridae) pupae and *Musca domestica* (Diptera: Muscidae) larvae. Biocontrol Science and Technology 11, 331-337.

Zanuncio, J.C., Molina-Rugama, A.J., Santos, G.P. and Ramalho, F.S. 2002. Effect of body weigth on fecundity and longevity of the stinkbug predator *Podisus rostralis*. Pesquisa Agropecuària Brasileira 37, 1225-1230.

Zanuncio, J.C., Silva, C.A.D., Lima, E.R., Pereira, F.F., Ramalho, F.S. and Serrlo, J.E. 2008. Predation rate of *Spodoptera frugiperda* (Lepidoptera: Noctuidae) larvae with and without defense by *Podisus nigrispinus* (Heteroptera: Pentatomidae). Brazilian Archives of Biology and Technology 51, 121-125.

CHAPTER II

Intra and interspecific interactions in the development and reproduction of *Podisus nigrispinus* and *Supputius cincticeps* (Heteroptera: Pentatomidae)

INTRODUCTION

The potential of Pentatomidae predators for biological control has been demonstrated in agricultural and forest ecosystems (Medeiros *et al.,* 2003a). *Podisus nigrispinus* (Dallas) and *Supputius cincticeps* (Stâl) (Heteroptera: Pentatomidae) are generalist predators from the Neotropical region of South America and consume a wide variety of prey, mainly insects from the orders Lepidoptera, Coleoptera, Diptera and Hemiptera (Lemos *et al.,* 2003). The exploitation of resources by predators is not restricted to the lower trophic level, as many can feed on each other in "intraguild predation" (Harvey *et al.*, 2011).

Functional characteristics that indicate whether a species is a good biological control agent, such as its efficient foraging ability and high predatory activity, also indicate whether the species can be an efficient intraguild predator (Alhmedi *et al.*, 2010). *Harmonia axyridis* (Pallas) (Coleoptera: Coccinellidae) feeds mainly on aphids and is a generalist predator that also feeds on other herbivores, predators and members of the same species (Snyder and Ives, 2003). This natural enemy is a good biological control agent and a dominant intraguild predator (Sato and Dixon, 2004; Alhmedi *et al.*, 2010). The decline in species associated with *H. axyridis* is mainly due to competition for resources (Michaud, 2002) and the fact that this predator is an aggressive intraguild predator (Alhmedi *et al.*, 2010). Some behavioral mechanisms can help prevent intraguild predation (Pallini *et al.*, 1998; Schellhorn and Andow, 1999), which depends on the abundance of prey and the developmental stages of the predators (Sato *et al.*, 2009).

Cannibalism is defined as the act of feeding on individuals of the same species and occurs in several insect species (Fox, 1975), especially with scarce food resources (Rickers and Scheu, 2005). Cannibalism in *P. nigrispinus* and *S. cincticeps* has been observed (Ramalho *et al.,* 2008; Pires *et al.*, 2011; personal observations), suggesting that intra- and interspecific competition can affect interactions between these species (Sato *et al.*, 2009).

Intra- and interspecific interactions between generalist predators according to biotic and abiotic variables are difficult to predict (Symondson *et al.,* 2002). These interactions can be measured by the influence of diet on the predator's life cycle (Sato *et al.*, 2009; Ware *et al.*, 2009), which can alter the life cycle of predatory pentatomids with better-fed individuals reproducing better and more frequently (Lemos *et al.*, 2001), as well as having shorter nymphal development (Zanuncio *et al.*, 2001). The secretion of juvenile hormones decreases under unfavorable conditions, preventing oogenesis and, consequently, egg development (Malaquias *et al.*, 2010). In addition, these conditions also hinder the development of the ovary (Lemos *et al.*, 2010). The size of females and the number of eggs of *Podisus maculiventris* (Say) (Heteroptera: Pentatomidae) were positively correlated (De Clercq *et al.*, 1998), and the number of spawns, eggs and nymphs were negatively affected by the lower weight of females *of*

Podisus rostralis (Stal) (Heteroptera: Pentatomidae) (Zanuncio *et al.*, 2002). Reproduction is reduced and nutrients allocated to survival when food is scarce (Vivian *et al.*, 2003). The growth rate of predator populations depends on the length of the stages, the survival rate per stage and the fecundity of the adults (Medeiros *et al.*, 2003b).

The aim of this study was to evaluate the impact of intra- and interspecies interactions and the occurrence of cannibalism and intraguild predation on the development and survival of nymphs and reproductive aspects of *P. nigrispinus* and *S. cincticeps*.

MATERIAL AND METHODS

The experiments were carried out at the Insect Biological Control Laboratory of the Institute of Biotechnology Applied to Agriculture (BIOAGRO) of the Federal University of Viçosa in Viçosa, Minas Gerais, Brazil.

Breeding *Podisus nigrispinus and Supputius cincticeps* (Heteroptera: Pentatomidae)

The predatory bedbugs were obtained from the massai brood of the Insect Biological Control Laboratory of the Institute of Biotechnology Applied to Agriculture (BIOAGRO) of the Federal University of Viçosa in Viçosa, Minas Gerais, Brazil.

Adults of these predators were placed in wooden cages (30 x 30 x 30 cm), with shingled sides and a glass lid, with pupae of *Tenebrio molitor* L. (Coleptera: Tenebrionidae) as food and moistened cotton as a source of water. The eggs of these insects were collected every two days and placed in Petri dishes (9.0 x 1.5 cm). Nymphs of these predators were reared in these Petri dishes until the fourth stage with *T. molitor* pupae and moistened cotton as a water source. After reaching the fourth stage, these nymphs were placed in wooden cages (30 x 30 x 30 cm) with shingled sides and glass lids, and *T. molitor* pupae and moistened cotton as a water source. The newly emerged adults were transferred to the adult cages. All rearing took place in an air-conditioned room at 25 ± 2°C, relative humidity of 60 ± 10% and 12-hour photophase (Zanuncio *et al.*, 2001).

Setting up the experiment

The experiment was carried out in the Insect Biological Control Laboratory in 500 mL plastic jars with a hole in the lid sealed with organza to increase aeration and a plastic tube of dental anesthetic with 2.5 ml of distilled water, sealed with cotton, to supply water and humidity (Lemos *et al.*, 2003). The jars were kept at 25 ± 1°C, 70 ± 10% relative humidity and 12-hour photophase in a randomized block design with two factors in a 3 x 2 scheme and fifteen replications.

One factor was species and the number of bed bug nymphs with three levels: Tl- fourteen second-stage nymphs of *P. nigrispinus;* T2- fourteen second-stage nymphs of *S, cincticeps;* T3- seven second-stage nymphs of each species. First-stage insects were not used because they had no predatory habit.

The other factor was the diet offered to the predators at two levels: Dl- *T, molitor pupae ad libitum*; D2- *T, molitor* pupae offered every four days. In the D2 diet, the pupae were removed 24 hours after being offered, even if they had not been predated, in order to avoid feeding outside the planned interval.

After the emergence of the adults, the bedbugs were sexed according to the external morphology of the genitalia and the size and shape of the body, and the sex ratio was calculated using the formula RS=number of females/number of males + females. The insects were randomly grouped into pairs and placed in 500 mL plastic jars with a hole in the lid sealed with organza to increase aeration and a plastic tube containing 2.5 ml of distilled water for the predators, with fifteen pairs per treatment. Each couple was exposed to the treatment in which their nymphs developed.

The eggs of the predators were collected and counted daily and placed in Petri dishes (9.0 x 1.2 cm) with a cotton pad soaked in distilled water and the nymphs counted 24 hours after hatching (Ramalho *et al,* 2008).

Statistical analysis

Survival, life cycle (second stage to adult emergence), weight, sex ratio and longevity of adults, numbers of eggs and nymphs, length of pre-oviposition, oviposition and post-oviposition periods, egg viability, egg incubation period, number of eggs per egg-laying and nymphs per egg-laying, total number of egg-layings, oviposition rate and nymph hatching rate per female of the predators and intraguild predation and cannibalism were evaluated. The data was submitted to analysis of variance using the F test and the means were compared using the Tukey test at a 5% significance level using the "R" statistical program (R Development Core Team, 2010).

RESULTS

Podisus nigrispinus development

The life cycle and duration of the stages of *P. nigrispinus,* with pupae of *T. molitor ad libitum*, alone or with *S. cincticeps*, was similar. However, the life cycle (F= 37.173; P= 2.913 x 10^{-13}) and the duration of the third (F= 23.048; P= 8.585 x 10-10), fourth (F= 24.1; P= 4.345 x 10-^{10}) and fifth stages (F= 27.677; P= 4.822 x 10-^{11}) were longer with feeding every four days. The length of the second stage was similar between treatments (F= 2.5696; P= 0.06353) (Table 1).

Table 1 - Means ± standard error[1] of the duration in days of each stage and of the life cycle (second stage to adult emergence) of *Podisus nigrispinus* (Heteroptera: Pentatomidae) at 25 ± 1°C, 70 ± 10% relative humidity and 12-hour photophase. Treatments: PNXD= *P. nigrispinus*, alone, with pupae of *Tenebrio molitor* (Coleoptera: Tenebrionidae) *ad libitum*; PNSCXD= *P. nigrispinus* with *Supputius cincticeps* (Heteroptera: Pentatomidae) and pupae of *T. molitor pupae ad libitum*; PNX4D= *P. nigrispinus*, alone, with *T. molitor* pupae every four days; PNSCX4D= *P. nigrispinus* with *S. cincticeps* and *T. molitor* pupae every four days.

Treatment	Second	Third	Bedroom	Fifth	Life cycle
PNXD	3,98 ± 0,12a	2,86 ± 0,11a	3,00 ± 0,05a	4,32 ± 0,07a	14,13 ± 0,26a
PNSCXD	3,86 ± 0,11a	2,77 ± 0,14a	3,17 ± 0,13a	4,42 ± 0,07a	14,19 ± 0,25a
PNX4D	3,84 ± 0,19a	3,55 ± 0,08b	4,08 ± 0,08b	5,93 ± 0,25b	17,04 ± 0,28b
PNSCX4D	3,49 ± 0,10a	3,82 ± 0,09b	4,00 ± 0,16b	5,84 ± 0,21b	16,86 ± 0,26b

(1) Averages followed by the same lower-case letter per column do not differ according to the "F" test at 5% probability.

The mortality of *P. nigrispinus nymphs*, alone or with *S. cincticeps*, with *T. molitor* pupae *ad libitum*, was similar. There was also no difference between these treatments with feeding every four days. However, nymph mortality was higher with feeding every four days than *ad libitum* (F= 22.536; P= 1.203 x 10^{-9}) (Table 2).

Table 2 - Means ± standard error[1] of the weight of females and males, mortality and sex ratio of *Podisus nigrispinus* (Heteroptera: Pentatomidae) at 25 ± 1°C, 70 ± 10% relative humidity and 12-hour photophase. Treatments: PNXD= *P. nigrispinus*, alone, with pupae of *Tenebrio molitor* (Coleoptera: Tenebrionidae) *ad libitum*; PNSCXD= *P. nigrispinus* with *Supputius cincticeps* (Heteroptera: Pentatomidae) and pupae of *T. molitor ad libitum*; PNX4D= *P. nigrispinus*, alone, with *T. molitor* pupae every four days; PNSCX4D= *P. nigrispinus* with *S. cincticeps* (Heteroptera: Pentatomidae) and *T. molitor* pupae every four days.

	Weight			
Treatment	Females	Males	Mortality (%)	Sexual reason
PNXD	72,86 ± 1,84a	54,24 ± 0,86a	6,19 ± 2,40a	0,44 ± 0,03a
PNSCXD	68,27 ± 1,61a	47,95 ± 1,06a	10,44 ± 3,23a	0,50 ± 0.05a
PNX4D	56,96 ± 1,74b	40,54 ± 1,34b	46,94 ± 5,69b	0,45 ± 0,05a
PNSCX4D	58,59 ± 2,29b	40,56 ± 1,52b	39,05 ± 5,12b	0,54 ± 0,08a

(1) Averages followed by the same lower-case letter per column do not differ according to the "F" test at 5% probability.

The weight of newly-merged females and males and the sex ratio (SR=number of females/number of males + females) of *P. nigrispinus*, alone or with *S. cincticeps*, with *T. molitor* pupae *ad libitum* did not differ between treatments. The same was true for treatments with *T. molitor* pupae every four days. However, the weight of adult females was lower with feeding every four days than *ad libitum* (F= 16.398; P= 1.156 x 10^{-7}), similarly to the weight of adult males (F= 31.424; P= 9.93 x 10^{-12}). The sex ratio was similar between treatments (F= 0.6782; P= 0.5691) (Table 2).

Development of *Supputius cincticeps*

The duration of the third (F= 18.023; P= 3.439 x 10^{-8}) and fourth (F= 11.98; P= 4.469 x 10^{-6}) stages of *S. cincticeps*, alone or with *P. nigrispinus*, with *T. molitor* pupae *ad libitum*, was shorter with *P. nigrispinus*. The other stages and the life cycle were similar between treatments. The duration of the third (F= 7.9993; P= 6.591 x 10^{-3}), fourth (F= 11.98; P= 4.469 x 10^{-6}) and fifth (F= 1.2426; P= 0.2706) stages and of the life cycle (F= 4.7818; P= 5.462 x 10^{-} 3) of *S. cincticeps, alone or with P. nigrispinus, was similar between treatments. cincticeps*, alone or with *P. nigrispinus*, with *T. molitor* pupae every four days was shorter with *P. nigrispinus. The* length of the second stage was similar between treatments (F= 0.887; P=0.4536) (Table 3).

Table 3 - Means ± standard error[1] of the duration in days of each stage and of the life cycle (second stage to adult emergence) of *Supputius cincticeps* (Heteroptera: Pentatomidae) at 25 ± 1°C, 70 ± 10% relative humidity and 12-hour photophase. Treatments: SCXD= *S. cincticeps*, alone, with pupae *of Tenebrio molitor* (Coleoptera: Tenebrionidae) *ad libitum*; SCPNXD= *S. cincticeps* with *Podisus nigrispinus* (Heteroptera: Pentatomidae) and pupae of *T. molitor pupae ad libitum*; SCX4D= *S. cincticeps*, alone, with *T. molitor* pupae every four days; SCPNX4D= *S. cincticeps* with *P. nigrispinus* and *T. molitor* pupae every four days.

Treatment	Second	Third	Bedroom	Fifth	Life cycle
SCXD	4,34 ± 0,15a	4,32 ± 0,21b	5,82 ± 0,35b	7,04 ± 0,14a	21,22 ± 0,41a
SCPNXD	4,83 ± 0,19a	2,77 ± 0,14a	4,35 ± 0,15a	6,44 ± 0,23a	19,05 ± 0,33a
SCX4D	4,71 ± 0,31a	5,79 ± 0,45c	6,27 ± 0,49b	8,99 ± 0,69b	24,05 ± 1,34b
SCPNX4D	4,75 ± 0,25a	4,59 ± 0,33b	4,16 ± 0,14a	7,21 ± 0,87a	20,93 ± 1,44a

(1) Averages followed by the same lower-case letter per column do not differ according to the "F" test at 5% probability.

The mortality of *S. cincticeps nymphs*, alone or with *P. nigrispinus*, with *T. molitor* pupae *ad libitum*, was similar. Mortality was also similar between treatments with feeding every four days. However, nymph mortality was higher with feeding every four days than with *ad libitum* feeding (F= 9.4234; P= 3.883 x 10^{-5}) (Table 4).

Table 4 - Means ± standard error[1] of the weight of females and males, mortality and sex ratio of *Supputius cincticeps* (Heteroptera: Pentatomidae) at 25 ± 1°C, 70 ± 10% relative humidity and 12-hour photophase. Treatments: SCXD= *S. cincticeps*, alone, with pupae *of Tenebrio molitor* (Coleoptera: Tenebrionidae) *ad libitum*; SCPNXD= *S. cincticeps* with *Podisus nigrispinus* (Heteroptera: Pentatomidae) and pupae of *T. molitor pupae ad libitum*; SCX4D = *S. cincticeps*, alone, with *T. molitor* pupae every four days; SCPNX4D= *S. cincticeps* with *P. nigrispinus* and *T. molitor* pupae every four days.

	Weight			
Treatment	Females	Males	Mortality (%)	Sexual reason
SCXD	42,61 ± 1,60b	28,44 ± 0,91a	59,89 ± 5,45a	0,51 ± 0,07a
SCPNXD	48,96 ± 1,86a	29,23 ± 1,20a	49,33 ± 6,83a	0,54 ± 0,06a
SCX4D	39,00 ± 2,04b	27,60 ± 1,01a	86,67 ± 3,87b	0,49 ± 0,12a
SCPNX4D	49,21 ± 5,85a	31,66 ± 1,50a	78,09 ± 5,55b	0,46 ± 0,13a

(1) Averages followed by the same lower-case letter per column do not differ according to the "F" test at 5% probability.

The weight of *S. cincticeps* females, alone or with *P. nigrispinus*, with *T. molitor* pupae *ad libitum* was higher for *S. cincticeps* with *P. nigrispinus* (F= 3.5164; P= 0.02439). The same occurred with feeding every four days. The weight of these females alone was similar (F= 1.0013; P= 0.3233) and lower than with *P. nigrispinus*, whose weights were also similar (F= 0.0043; P= 0.9483). Male weight (F= 1.6279; P= 0.1985) and sex ratio (F=0.1393; P= 0.936) were similar between treatments (Table 4).

Cannibalism and intraguild predation

Cannibalism and intraguild predation occurred only with feeding every four days, mainly on individuals in ecdysis (Figure 1, 2), with the two species together. Intraguild predation was bidirectional, i.e. nymphs of *P. nigrispinus* preyed on nymphs *of S cincticeps*, and the other way around.

Figura 1 - Cannibalism between nymphs *of Podisus nigrispinus* (figures A and B) and *Supputius cincticeps* (Heteroptera: Pentatomidae) (figures C and D).

Figura 2 - Bidirectional intraguild predation between *Podisus nigrispinus* and *Supputius cincticeps* (Heteroptera: Pentatomidae). A and B- *P. nigrispinus* nymphs preying on *S. cincticeps* nymphs; C and D- *S. cincticeps* nymphs preying on *P. nigrispinus* nymphs.

Reproductive aspects of *Podisus nigrispinus*

The reproductive aspects were similar between treatments with *ad libitum* feeding *and* better than every four days, which were also similar.

The pre-oviposition period was longer with feeding every four days (F= 10.026; P= 2.357×10^{-5}). The oviposition (F= 0.7679; P= 0.517) and post-oviposition (F= 1.0027; P= 0.3988) periods were similar between treatments (Table 5). Table 5 - Means ± standard error[1] of the reproductive parameters of *Podisus nigrispinus* (Heteroptera: Pentatomidae) at 25 ± 1°C, 70 ± 10% relative humidity and 12-hour photophase. Treatments: PNXD= *P. nigrispinus*, alone, with pupae of *Tenebrio molitor* (Coleoptera: Tenebrionidae) *ad libitum*; PNSCXD= *P. nigrispinus* with *Supputius cincticeps* (Heteroptera: Pentatomidae) and pupae of *T. molitor pupae ad libitum*; PNX4= *P. nigrispinus*, alone, with *T. molitor* pupae every four days; PNSCX4D= *P. nigrispinus* with *S. cincticeps* and *T. molitor* pupae every four days.

	Treatments			
Reproductive parameters	PNXD	PNSCXD	PNX4D	PNSCX4D
Number of eggs per female	577,27 ± 64,03a	464,67 ± 54,57a	287,33 ± 65,95b	239,53 ± 54,53b
Number of eggs/posture	24,67 ± 0,86a	24,07 ± 0,89a	14,75 ± 1,58b	15,84 ± 1,55b
Number of spawnings per female	23,80 ± 2,86a	20,00 ± 2,65a	18,06 ± 3,23a	14,40 ± 2,88a
Pre-oviposition (days)	4,87 ± 0,40a	4,73 ± 0,32a	12,93 ± 2,38b	11,21 ± 1,34b
Oviposition (days)	37,47 ± 5,19a	29,13 ± 4,19a	40,21 ± 6,27a	36,71 ± 6,11a
Post-oviposition (days)	4,20 ± 0,91a	2,53 ± 1,02a	6,57 ± 2,96a	3,86 ± 1,05a
Egg incubation (days)	5,07 ± 0,05a	5,06 ± 0,05a	5,23 ± 0,08a	5,20 ± 0,11a
Egg viability (%)	85,33 ± 3,02a	84,98 ± 4,29a	69,34 ± 5,32b	62,16 ± 6,63b
Number of nymphs per female	494,40 ± 52,49a	409,27 ± 52,84a	215,13 ± 48,86b	167,80 ± 45,84b
Number of nymphs/post	21,62 ± 1,14a	21,10 ± 1,45a	10,85 ± 1,49b	9,85 ± 1,54b
Longevity of females (days)	46,53 ± 5,26a	37,07 ±	56,80 ± 6,06a	48,93 ± 7,04a

		4,28a		
Longevity of males (days)	43,93 ± 5,09a	40,27 ± 6,00a	46,93 ± 5,71a	51,20 ± 6,85a

(1) Averages followed by the same lower-case letter per line do not differ according to the "F" test at 5% probability.

The number of spawnings per female (F= 1.8063; P= 0.1565) and the egg incubation period (F= 1.4177; P= 0.2479) were similar between treatments. The number of eggs (F= 6.8334; P= 5.269 X 10^{-4}) and nymphs per female (F= 9.6274; P= 3.191 X 10^{-5}) and eggs (F= 17.919; P= 3.384 X 10^{-8}) and nymphs (F= 20.495; P= 5.374 X 10^{-9}) per laying were lower with feeding every four days. Male longevity (F= 0.6068; P= 0.6134) was similar between treatments and the same was true for females (F= 1.9979; P= 0.1247) (Table 5).

Egg viability (F= 5.5315; P= 2.192 X 10^{-3}) was lower with feeding every four days (Table 5).

Reproductive aspects of *Supputius cincticeps*

The high mortality of *S. cincticeps n*ymphs fed every four days made it impossible to assemble fifteen pairs in these treatments, which had seven and eight pairs with *S. cincticeps* alone or with *P. nigrispinus*, respectively. The reproductive aspects differed between treatments and isolated females fed *ad libitum* had the best reproductive performance.

Oviposition (F= 1.5069; P= 0.2303), post-oviposition (F= 0.0857; P= 0.9674) and egg incubation (F= 1.521; P= 0.0789) periods were similar between treatments. The pre-oviposition period was longer (F= 3.4463; P= 2.73 x 10^{-2}) and the number of eggs (F= 4.7847; P= 5.94 x 10^{-3}) and spawns per female (F= 4.4949; P= 8.117 X 10^{-3}) shorter with feeding every four days. The number of eggs per lay was lower (F= 3.6542; P= 2.192 X 10^{-2}) for *S. cincticeps* in isolation and fed every four days.

Egg viability was higher with isolated *S. cincticeps* (F= 6.9631; P= 8.905 x 10^{-4}). The number of nymphs per female (F= 3.9887; P= 1.395 x 10^{-2}) and per lay (F= 4.3448;

P= 1.074 x 10^{-2}) was higher for *S. cincticeps*, isolated and fed *ad libitum* (Table 6).

The longevity of females (F= 1.703; P= 0.1814) and males (F= 0.6367; P= 0.5957) was similar between treatments (Table 6).

Table 6 - Means ± standard error[1] of the reproductive parameters of *Supputius cincticeps* (Heteroptera: Pentatomidae) at 25 ± 1°C, 70 ± 10% relative humidity and 12-hour photophase. Treatments: SCXD= *S. cincticeps*, alone, with pupae *of Tenebrio molitor* (Coleoptera: Tenebrionidae) *ad libitum*; SCPNXD= *S. cincticeps* with *Podisus nigrispinus* (Heteroptera: Pentatomidae) and pupae of *T. molitor pupae ad libitum*; SCX4D= *S. cincticeps*, alone, with *T. molitor* pupae every four days; SCPNX4D= *S. cincticeps* with *P. nigrispinus* and *T. molitor* pupae every four days.

	Treatments			
Reproductive parameters	SCXD	SCPNXD	SCX4D	SCPNX4D
Number of eggs per female	126,80 ± 2,20a	95,07 ± 1,72a	24,38 ± 1,54b	16,57 ± 0,43b
Number of eggs/posture	10,62 ± 0,58a	12,34 ± 0,94a	7,32 ± 0,78b	10,07 ± 1,82a
Number of spawnings per female	11,07 ± 2,20a	7,87 ± 1,72a	3,13 ± 1,54b	1,57 ± 0,43b
Pre-oviposition (days)	9,36 ± 0,91a	10,21 ± 1,50a	17,00 ± 2,21b	14,60 ± 3,75b
Oviposition (days)	19,64 ± 3,51a	18,43 ± 3,28a	13,20 ± 7,06a	7,00 ± 1,14a
Post-oviposition (days)	3,50 ± 1,07a	3,36 ± 1,28a	4,40 ± 2,29a	4,00 ± 1,37a
Egg incubation (days)	6,20 ± 0,06a	5,95 ± 0,08a	6,37 ± 0,27a	6,00 ± 0,00a
Egg viability (%)	65,89 ± 6,61a	35,30 ± 9,61b	70,85 ± 4,61a	11,17 ± 7,28b
Number of nymphs per female	87,47 ± 19,89a	45,80 ± 18,30b	16,88 ± 8,14c	2,14 ± 1,70d
Number of nymphs/post	7,10 ± 0,81a	4,19 ± 1,15b	5,21 ± 0,75b	1,10 ± 0,78c
Longevity of females (days)	32,00 ± 3,21a	31,73 ± 2,87a	26,88 ± 5,48a	20,57 ± 3,99a
Longevity of males (days)	22,00 ± 3,65a	21,60 ± 3,67a	19,00 ± 4,00a	14,14 ± 4,68a

(1) Averages followed by the same lower-case letter per line do not differ according to the "F" test at 5% probability.

DISCUSSION

Podisus nigrispinus development

The duration of the second stage to adult and of the stages were similar for *P. nigrispinus* alone or with *S. cincticeps*, with *T. molitor* pupae *ad libitum* or every four days, showing that interspecific competition did not harm the development of this predator. The longer life cycle and stage length of *P. nigrispinus* with *T. molitor* larvae (Lemos *et al*., 2003) may be due to the fact that these authors used larvae of the prey and individualized nymphs, which increases the energy expenditure to dominate the prey, since different nymphs of this predator can attack the same prey, reducing the energy cost. The longer life cycle and duration of the third, fourth and fifth stages of *P. nigrispinus*, with pupae every four days, shows that this predator has acquired less nutrients to increase its size and carry out ecdysis. Longer nymphal periods consequently lead to later reproduction and fewer generations per year. The similar length of the second stage between treatments is related to the small amount of food needed to change instar at this stage (Santos *et al*., 1996).

The lower mortality of *P. nigrispinus* nymphs with *T. molitor* pupae *ad libitum* shows the absence of cannibalism and intraguild predation and the maintenance of the nymphs' vital needs, and agrees with the mortality of around 10% of *P. nigrispinus* with *T. molitor* larvae (Torres *et al*., 1998). Cannibalism and intraguild predation with pupae every four days increased the mortality of bed bug nymphs, with similar mortality values and a similar effect of intra- and interspecific interactions.

The similar weight of newly emerged females *of P. nigrispinus* alone or with *S. cincticeps*, with *T. molitor* pupae *ad libitum* or every four days, shows that interspecific competition did not affect the weight of this predator. The same happened with the weight of the males. *P. nigrispinus* females weighing more than 60 mg with *T. molitor* pupae *ad libitum* is important because it leads to better reproductive performance and an increase in the predator's population in the field (Zanuncio *et al*. 2002). On the other hand, the lower body mass of females of this predator with pupae every four days may indicate undernourishment, as reported for the weight of adults of this predator with increasing feeding interval (Oliveira *et al.*, 2002a). The greater weight of *P. nigrispinus* males alone with pupae *ad libitium* is not important for reproduction, as this parameter does not benefit the reproductive aspects of their females (Rodrigues *et al.,* 2009). The similar sex ratio was due to similar mortality between females and males in all treatments.

Development of *Supputius cincticeps*

The similar length of the second stage to adult of *S cincticeps* alone or with *P. nigrispinus*, and pupae *ad libitum*, shows that interspecific competition did not harm the development of this predator, and is similar to that of this predator with *T. molitor* larvae (Beserra *et al.*, 1995). The shorter duration of the third and fourth stages of this predator with *P. nigrispinus* may be due to increased feeding due to competition (Relyea, 2004). The similar duration of the second stage between treatments is related to the small amount of food needed to change instar at this stage. The similar duration of the fifth stage is due to the need to accumulate biomass for the reproductive system to start growing (Oliveira *et al.*, 2004). The shorter life cycle of *S. cincticeps* with *P. nigrispinus* with pupae every four days shows that the interaction of this predator with *P. nigrispinus* is beneficial, as it can lead to an increase in the number of generations per year. The nymphal development of *P. nigrispinus* at all stages was shorter than that of *S. cincticeps*, indicating that the latter can take advantage of the ecdysis moments of *P. nigrispinus* to carry out intraguild predation and obtain energy to complete its life cycle more quickly. The pressure of competition and intraguild predation may lead this predator to reallocate more resources to reduce the nymphal period, as the relative size of predators affects predation success (Straub *et al.*, 2008; Hogg and Daane, 2011). In addition, predators with a shorter nymphal period can reproduce earlier and have a greater number of generations per year (Zanuncio *et al.*, 2001).

The mortality of *S. cincticeps nymphs* with *T. molitor* pupae *ad libitum* was higher than that of *P. nigrispinus*, showing that some predators are more adapted to certain prey, with which they develop faster and survive longer (Waddil and Shepard, 1974). Offering pupae every four days increased the mortality of this bug's nymphs, due to undernourishment, cannibalism and intraguild predation.

The weight of *S. cincticeps* females alone was similar to that found with *T. molitor* larvae (Zanuncio *et al.*, 2005). The greater weight of *S. cincticeps* females with *P. nigrispinus* shows the importance of this interspecific interaction, as this parameter has a positive correlation with the fecundity of female predatory bedbugs (Oliveira *et al.*, 2005). Pupae *of T. molitor* are large and, for this reason, nymphs of *P. nigrispinus* and *S. cincticeps* can share the same prey. Nymphs *of S. cincticeps* may wait for those of *P. nigrispinus* to dominate the prey and then consume it, thus saving energy. In addition, digestion in these species is extraoral (Azevedo *et al.,* 2007) and *S cincticeps* may take advantage of enzymes or food already digested by *P. nigrispinus* to feed, which would also lead to energy savings, as observed for spiders (Amir *et al.*, 2000). The pressure of competition and intraguild predation may lead this predator to reallocate more resources to higher growth, as the relative size of predators affects the outcome of predation (Straub *et al.*, 2008; Hogg and Daane, 2011). The weight of *S. cincticeps* males in all treatments was similar to that of this predator with *T. molitor* larvae (Zanuncio *et al.*, 2005). The importance of the male in reproduction depends on each species, as the number of fertile eggs of *Nezara viridula* (Linnaeus) (Heteroptera: Pentatomidae) females was higher with males of greater body mass (McClain *et al.* 1991). However, the weight of *P. nigrispinus* males does not benefit the reproductive aspects of their females (Rodrigues *et al.*, 2009). The similar sex ratio was due to similar mortality between females and males in all treatments.

Cannibalism and intraguild predation

Cannibalism and intraguild predation occurred between *P. nigrispinus* and *S. cincticeps* with lethal responses to both predators without shared prey. Alternative prey, such as intraguild and co-species predation can maintain the survival of these predators while the essential prey (shared prey) is not present (Evans, 2008). The ability of a species to persist during a shortage of essential prey depends on its acceptance of alternative foods (Ware *et al*., 2009). Nymphs of *Brontocoris tabidus* (Signoret) (Heteroptera: Pentatomidae) and *P. nigrispinus* prey on close co-species in the absence of food (Pires *et al.,* 2011). This was also observed for *H. axyridis* and *Coccinella septempunctata brucki* Mulsant (Coleoptera: Coccinellidae) preying on fewer aphids, indicating that prey abundance determines the incidence of intraguild predation and/or cannibalism (Sato *et al*., 2009; Chacón and Heimpel, 2010). Cannibalism and intraguild predation can occur even when food is abundant (Chapman *et al*., 1999; Mallampalli *et al*., 2002; De Clercq *et al*., 2003; Bonte and De Clercq, 2010), but this was not observed. *Podisus nigrispinus* and *S. cincticeps* are similar in size and mobility and both are generalists. This is why they show bidirectional intraguild predation, which is directly linked to body size, mobility and food specificity (Lucas, 2005).

The low defense of the predator during ecdysis explains the higher rates of cannibalism and intraguild predation during this phase (Sato and Dixon, 2004), when the formation of the new integument occurs in a process called "*tanning*", with darkening due to melanization and stiffening due to the sclerotization of the new cuticle (Elias-Neto *et al.,* 2009). At this stage, the insects are defenseless due to the immobility and low resistance of the integument and are easy prey for cannibalistic individuals and other predators. Cannibalism and intraguild predation can increase the mortality rate, but are favorable as they increase the chances of survivors dispersing (Ruberson *et al*., 1986; Kudo and Nakahira, 2005; Ohba *et al.,* 2006). In addition, they can stabilize ecological relationships and regulate insect population density by reducing competition for resources and ensuring survival (Richardson *et al.,* 2010).

The high rates of cannibalism and intraguild predation in *P. nigrispinus* and *S. cincticeps*, with less food, may be due to space limitations and a high density of predators in the pots, which increases these interactions (Bonte and De Clercq, 2010). Insects with high rates of co-specific predation tend to be highly dispersed, which reduces this habit (Ichikawa and Kurauchi, 2009). The dispersal of individuals reduces the effect of local density and competition for resources, reducing the chances of intra-

and interspecific interactions (Rudolf *et al.*, 2010). Predatory bedbugs of the *Podisus* genus have dispersed populations in the field (Sant'Ana *et al.*, 1997), but the limited environment in the pots can increase rates of cannibalism and intraguild predation. The presence of plants can reduce cannibalism (Sinia *et al.*, 2004) and intraguild predation (Hosseini *et al.*, 2010). This is because Asopinae are zoophytophagous predators (Zanuncio *et al.*, 2004; Fialho *et al.*, 2009) and can be favored by phytophagy during food shortages (Oliveira *et al.*, 2002b).

Reproductive aspects of *Podisus nigrispinus*

The reproductive aspects of *P. nigrispinus* show that the interspecific interaction with *S'. cincticeps* did not affect these parameters. The diet factor was more important, as females fed every four days had worse reproductive performance. However, the four-day feeding interval did not affect the reproduction of *P. nigrispinus* (Molina-Rugama *et al.*, 1998). This can be explained by the fact that these authors only analyzed the feeding interval in the adult stage, which shows the importance of frequent feeding in the nymphal stage.

The pre-oviposition period with *ibitium* feeding was similar to that of other studies with *P. nigrispinus* (Ramalho *et al.*, 2008; Malaquias *et al.*, 2010; Holtz *et al.*, 2011). However, this period was longer with feeding every four days, as the females need to gain more weight before starting to reproduce (Zanuncio *et al.*, 2002). *Podisus rostralis* also showed a negative linear correlation between the weight of its females and the pre-oviposition period (Zanuncio *et al.*, 2002).

The lower number of eggs and nymphs is due to the weight of their females below 60 mg with food every four days. This was demonstrated for females *of P. nigrispinus* (Dallas) (Heteroptera: Pentatomidae) weighing between 61 and 70 mg, which had a higher number of eggs per lay and nymphs per day (Espindula *et al., 2006),* and the number of eggs and nymphs was negatively affected by the lower weight of *P. rostralis* females (Zanuncio *et al.*, 2002). In addition, the quality and quantity of food during the immature and adult stages affect fertility (Medeiros *et al.*, 2003b) and ovary development, which decreases the egg production of *P. nigrispinus* (Lemos *et al.*,

2010). Males and females reduce reproductive effort to favor their longevity as observed for other Asopinae (Legaspi *et al.*, 1996; Molina-Rugama *et al.*, 2001; Mourao *et al.*, 2003), explaining the similar longevity between treatments. This phenomenon is known as a physiological trade-off and consists of the allocation of energy by an organism to two or more processes that compete directly for limited resources (Stearns, 1994). This also explains the lower number of eggs and nymphs feeding at four days. *Podisus maculiventris* maintained its weight and reproductive performance only with abundant prey (O'Neil and Wiedenmann, 1990).

The viability of eggs from females fed *ad libitum* was similar to that observed for *P. nigrispinus* fed daily on *Alabama argillacea* (Hübner) (Lepidoptera: Noctuidae) caterpillars (Medeiros *et al.*, 2003b). The lower viability with feeding every four days is due to the lower body mass of these insects with an inadequate diet, which leads to poor egg formation (Lemos *et al.*, 2009). In addition, the secretion of juvenile hormones can decrease, impairing oogenesis and, consequently, egg development (Malaquias *et al.*, 2010). Another factor that affected viability was egg predation by females fed every four days, which has already been observed by *P. nigrispinus* with scarce prey. This may be an additional adaptation of predators to increase their chances of survival during periods of prey scarcity (Ramalho *et al.*, 2008). Eggs, being immobile, are very vulnerable to cannibalism when prey is scarce (Agarwala, *et al.*, 2003)

Reproductive aspects of *Supputius cincticeps*

Reproductive aspects of *S, cincticeps* show the influence of interspecific interaction with *P, nigrispinus* and diet, as the reproductive performance of this predator was better when isolated and fed *ad libitum*.

Larger females should have had better reproductive performance, as observed for *P, nigrispinus, P, rostralis* and *S, cincticeps* (Mohaghegh *et al.*, 1999; Zanuncio *et al.*, 2002, 2005), with larger females allocating more energy to reproduction (Oliveira *et al.*, 2003). Larger females *of S. cincticeps* have a higher number of spawns, eggs and nymphs (Oliveira *et al.*, 2003). The pre-oviposition period of *P, rostralis* was longer for lighter females, as they needed to acquire more weight before starting to reproduce (Zanuncio *et al.,* 2002). Females *of S, cincticeps* that developed with *P, nigrispinus* and fed every four days were larger, but had worse reproductive aspects. This may be due to the feeding interval, as reported for *S, cincticeps* with *T, molitor* larvae at intervals of one, two, four, six or eight days. This predator showed lower percentages of spawn, eggs and nymphs as the prey-free interval increased (Mourao *et al*., 2003). Reducing the quality and quantity of food reduces the reproductive parameters of insects, which start to invest in survival (Vivan *et al,* 2002; Holtz *et al,* 2009).

Females *of S, cincticeps* that developed with *P, nigrispinus* and fed *ad libitum* were larger and had lower egg viability and, consequently, fewer nymphs. The possible explanation for this low viability, which also occurs for females fed *P. nigrispinus* every four days, is that *S. cincticeps nymphs* allocate more resources to faster development and the production of larger females, in order to reduce intraguild competition and predation, than to the development of reproductive structures, leading to poor egg formation and lower egg viability. This hypothesis is possible because the nutritional resources acquired during the immature stage are used for the development and formation of reproductive structures (Dossi and Cônsoli, 2010). In addition, energy resources can be reallocated due to the insect's needs, such as in the immune system (Schmid-Hempel, 2005), which leads to *trade-offs* and a decrease in *fitness* (Mikolajewski *et al.,* 2008). This was observed for the predator *Coccinella septempunctata* (L.) (Coleoptera: Coccinellidae), which produced a greater number of eggs per day with healthy aphids than with aphids infected with the entomopathogenic fungus *Neozygites fresenii* (Simelane *et al.*, 2008). The similar viability of the eggs in the treatments with this isolated species shows that the diet did not affect the development of the eggs and the absence of egg predation by females in the treatment with feeding every four days.

The similar longevity of males and females of *S. cincticeps* between treatments shows that pentatomids prioritize reproduction in situations of abundant food, but their reproductive parameters are compromised as food decreases in quality and quantity and the insects start to invest in survival (Vivan *et al.*, 2002; Holtz *et al.*, 2009). This has been reported for the similar longevity *of S. cincticeps* with *T. molitor* larvae at intervals of one, two, four, six or eight days (Mourao *et al.*, 2003). The longevity reported by this author is similar to that of this study. In addition, the weight of the female did not affect the longevity *of S. cincticeps*, as observed by this predator, where the longevity of females with different weights was similar (Oliveira *et al.*, 2003).

CONCLUSION

Intra- and interspecific interactions did not affect the development and reproduction of *P. nigrispinus*, with *ad libitum* feeding, but cannibalism and intraguild predation were recorded with feeding every four days.

Interspecific interactions improved the development of *S cincticeps*, with feeding *ad libitum* and every four days, but its reproduction was impaired. The population of this species must be kept at low densities in contact with *P. nigrispinus*. Cannibalism and intraguild predation occurred with feeding every four days and helped the survival of part of its nymphs until the main prey was located.

The conservation of both predators is interesting in conservative biological control programs, since their coexistence is possible and does not affect the main predator, *P. nigrispinus*. It is better to use only *P. nigrispinus* in inoculative biological control programs, due to the low viability of *S. cincticeps* eggs with this predator, and it is better to invest financial resources in producing more *P. nigrispinus* individuals.

REFERENCES

Agarwala, B.K., Bardhanroy, P., Yasuda, H. and Takizawa, T. 2003. Effects of conspecific and heterospecific competitors on feeding and oviposition of a predatory ladybird: a laboratory study. Entomologia Experimentalis et Applicata 106, 219-226.

Alhmedi, A., Haubruge, E. and Francis, F. 2010. Intraguild interactions and aphid predators: biological efficiency of *Harmonia axyridis* and *Episyrphus balteatus*. Journal of Applied Entomology 134, 34-44.

Amir, N., Whitehouse, M.E.A. and Lubin, Y. 2000. Food consumption rates and competition in a communally feeding social spider, *Stegodyphus dumicola* (Eresidae). Journal of Arachnology 28, 195-200.

Azevedo, D.O., Zanuncio, J.C., ZanuncioJunior, J.S., Martins, G.F., Marques- Silva, S., Sossai, M.F. and Serrao, J.E. 2007. Biochemical and morphological aspects of the predator *Brontocoris tabidus* (Heteroptera: Pentatomidae). Brazilian Archives of Biology and Technology 50, 469-477.

Beserra, E.B., Zanuncio, T.V., Zauncio, J.C. and Santos, G.P. 1995. Development of *Supputius cincticeps* (Heteroptera, Penatatomidae) fed with larvae of *Zophobas confusa, Tenebrio molitor* (Coleoptera: Tenebrionidae) and *Musca domestica* (Diptera, Muscidae). Revista Brasileira de Zoologia 12, 723-733.

Bonte, M. and De Clercq, P. 2011. Influence of predator density, diet and living substrate on developmental fitness of *Orius laevigatus*. Journal of Applied Entomology 135, 343-350.

Chacón, J.M. and Heimpel, G.E. 2010. Density-dependent intraguild predation of an aphid parasitoid. Oecologia 164, 213-220.

Chapman, J.W., Williams, T., Escribano, A., Caballero, P., Cave, R.D. and Goulson, D. 1999. Age-related cannibalism and horizontal transmission of a nuclear polyhedrosis cirus in larval *Spodoptera frugiperda*. Ecological Entomology 24, 268275.

De Clercq, P., Vandewalle, M. and Tirry, L. 1998. Impact of inbreeding on performance of the predator *Podisus maculiventris*. BioControl 43, 299-310.

De Clercq, P., Peeters, I., Vergauwe, G. and Thas, O. 2003. Interaction between *Podisus maculiventris* and *Harmonia axyridis,* two predators used in augmentative biological control in greenhouse crops. BioControl 48, 39-55.

Dossi, F.A.C. and Cônsoli, F.L. 2010. Ovarian development and crown influence on ovary maturation in *Diaphorina citri* Kuwayama (Hemiptera: Psyllidae). Neotropical Entomology 39, 414-419.

Elias-Neto, M., Soares, M.P.M. and Bitondi, M.M.G. 2009. Changes in integument structure during the imaginal molt of the honey bee. Apidologie 40, 2939.

Espindula, M.C., Oliveira, H.N., Campanharo, M., Pastori, P.L. and Magevski, G.C. 2006. Influence of body mass on reproductive characteristics and longevity of *Podisus nigrispinus* (Dallas) (Heteroptera: Pentatomidae) females. Idesia 24, 19-25.

Evans, E.W. 2008. Multitrophic interactions among plants, aphids, alternate prey and shared natural enemies - a review. European Journal of Entomology 105, 369-380.

Fialho, M.C.Q., Zanuncio, J.C., Neves, C.A., Ramalho, F.S. and Serrao, J.E. 2009. Ultrastructure of the digestive cells in the midgut of the predator *Brontocoris tabidus* (Heteroptera: Pentatomidae) after different feedings periods on prey and plants. Annals of the Entomological Society of America 102, 119-127.

Fox, L.R. 1975. Cannibalism in natural populations. Annual Review of Ecology, Evolution, and Systematics 6, 87-106.

Harvey, J.A., Pashalidou, F., Soler, R. and Bezemer, T.M. 2011. Intrinsic competition between two secondary hyperparasitoids results in temporal trophic switch. Oikos 120, 226-233.

Hogg, B.N. and Daane, K.M. 2011. Diversity and invasion within a predator community: impacts on herbivore suppression. Journal of Applied Ecology 48, 453461.

Holtz, A.M., Almeida, G.D., Fadini, M.A.M., Zanuncio-Junior, J.S., Zanuncio, T.V. and Zanuncio, J.C. 2009. Survival and reproduction of *Podisus nigrispinus* (Heteroptera: Pentatomidae): effects of prey scarcity and plant feeding. Chilean Journal of Agricultural Research 69, 468-472.

Holtz, A.M., Almeida, G.D., Fadini, M.A.M., Zanuncio-Junior, J.S., Zanuncio, J.C. and Andrade, G.S. 2011. Phytophagy on eucalyptus plants increases the development and reproduction of the predator *Podisus nigrispinus* (Hemiptera: Pentatomidae). Acta Scientiarum Agronomy 33, 231-235.

Hosseini, M., Ashouri, A., Enkegaard, A., Weisser, W.W., Goldansaz, H., Mahalati,

M.N. and Moayeri, H.R.S. 2010. Plant quality effects on intraguild predation between *Orius laevigatus* and *Aphidoletes aphidimyza*. Entomologia Experimentalis et Applicata 135, 208-216.

Ichikawa, T. and Kurauchit, T. 2009. Larval cannibalism and pupal defense against cannibalism in two species of tenebrionid beetles. Zoological Science 26, 525-529

Kudo, S. and Nakahira, T. 2005. Trophic-egg production in a subsocial bug: adaptive plasticity in response to resource conditions. Oikos 111, 459-464.

Legaspi, J.C. and Legaspi Jr., B.C. 2004. Does a polyphagous predator prefer prey species that confer reproductive advantage? Case study of *Podisus maculiventris*. Environmental Entomology 33, 1401-1409.

Lemos, W.P., Medeiros, R.S., Ramalho, F.S. and Zanuncio, J.C. 2001. Effects of plant feeding on the development, survival and reproduction of *Podisus nigrispinus* (Heteroptera: Pentatomidae). International Journal of Pest Management 47, 89-93.

Lemos, W.P., Ramalho, F.S., Serrao, J.E. and Zanuncio, J.C. 2003. Effects of diet on development of *Podisus nigrispinus* (Dallas) (Heteroptera: Pentatomidae), a predator of the cotton leafworm. Journal of Applied Entomology 127, 389-395.

Lemos, W.P., Zanuncio, J.C., Ramalho, F.S. and Serrao, J.E. 2009. Fat body of the zoophytophagous predator *Brontocoris trabidus* (Heteroptera: Pentatomidae) females: impact of the herbivory. Micron 40, 635-638.

Lemos, W.P., Zanuncio, J.C., Ramalho, F.S., Zanuncio, T.V. and Serrao, J.E. 2010. Herbivory affects ovarian development in the zoophytophagous predator *Brontocoris tabidus* (Heteroptera: Pentatomidae). Journal of Pest Science 83, 69-76.

Lucas, E. 2005. Intraguild predation among aphidophagous predators. European Journal of Entomology 102, 351-364.

Mallampalli, N., Castellanos, I. and Barbosa, P. 2002. Evidence for intraguild predation by *Podisus maculiventris* on a ladybeetle, *Coleomegilla maculata:* implications for biological control of Colorado potato beetle, *Leptinotarsa decemlineata.* Biocontrol 47, 387-398.

Malaquias, J.B., Ramalho, F.S., Fernandes, F.S., Nascimento Júnior, J.L., Correia, E.T. and Zanuncio, J.C. 2010. Effects of photoperiod on reproduction and longevity of *Podisus nigrispinus* (Heteroptera: Pentatomidae). Annals of Entomological Society of America 103, 603-610.

McClain, D.K. 1991. Heritability of size: a positive correlate of multiple fitness components in the Southern green stink bug (Hemiptera: Pentatomidae). Annals of the Entomological Society of America 84, 174-178.

Medeiros, R.S., Ramalho, F.S., Zanuncio, J.C. and Serrao, J.E. 2003a. Effect of temperature on life table parameters of *Podisus nigrispinus* (Heteroptera: Pentatomidae) fed with *Alabama argillacea* (Lepidoptera: Noctuidae) larvae. Journal of Applied Entomology 127, 209-213.

Medeiros, R.S., Ramalho, F.S., Serrao, J.E. and Zanuncio, J.C. 2003b. Temperature influence on the reproduction of *Podisus nigrispinus*, a predator of the noctuid larva *Alabama argillacea.* BioControl 48, 695-704.

Michaud J.P. 2002. Invasion of the Florida citrus ecosystem by *Harmonia axyridis* (Coleoptera: Coccinellidae) and asymmetric competition with a native species, *Cycloneda sanguinea.* Environmental Entomology 31, 827-835.

Mikolajewski, D.J., Stoks, R. and Joop, G. 2008. Predators and cannibals modulate sex-specific plasticity in life-history and immune traits. Functional Ecology 22, 114-120.

Mohaghegh, J., De Clercq, P. and Tirry, L. 1999. Effects of rearing history and geographical origin on reproduction and body size of the predator *Podisus nigrispinus* (Heteroptera: Pentatomidae). European Journal of Entomology 96, 6972.

Molina-Rugama, A.J., Zanuncio, J.C., Pratissoli, D. and Cruz, I. 1998. Effect of Feeding Interval on Reproduction and Longevity of the Predator *Podisus nigrispinus* (Dallas) (Heteroptera: Pentatomidae). Annals of the Entomological Society of Brazil 27, 77-84.

Molina-Rugama, A.J., Zanuncio, J.C., Vinha, E. and Ramalho, F.S. 2001. Daily rate of egg laying of the predator *Podisus rostralis* (Stal, 1860) (Heteroptera, Pentatomidae) under different feeding intervals. Revista Brasileira de Entomologia 45, 1-5.

Mourào, S.A., Zanuncio, J.C., Molina-Rugama, A.J., Vilela, E.F. and Lacerda M.C. 2003. Effect of prey scarcity on the survival and reproduction of the predator *Supputius cincticeps* (Stâl) (Heteroptera: Pentatomidae). Neotropical Entomology 32, 469-473.

O'Neil, R.J. and Wiedenmann, R.N. 1990. Body weight of *Podisus maculiventris* (Say) under various feeding regimes. Canadian Entomologist 122, 285294.

Ohba, S., Hidaka, K. and Sasaki, M. 2006. Notes on paternal care and sibling cannibalism in the giant water bug, *Lethocerus deyrolli* (Heteroptera: Belostomatidae). Entomological Science 9, 1-5.

Oliveira, H.N., Pratissoli, D., Pedruzzi, E.P. and Espindula, M.C. 2004. Development of the predator *Podisus nigrispinus* fed with *Spodoptera frugiperda and Tenebrio molitor.* Pesquisa Agropecuària Brasileira 39, 947-951.

Oliveira, I., Zanuncio, J.C., Serrao, J.E., Zanuncio, T.V., Pinon, T.B.M. and Fialho, M.C.Q. 2005. Effect of female weight on reproductive potential of the predator *Brontocoris tabidus* (Signoret, 1852) (Heteroptera: Pentatomidae). Brazilian Archives of Biology and Technology 48, 295-301.

Oliveira, J.E.M., Torres, J.B., Moreira, A.F.C. and Ramalho, F.S. 2002a. Biology of *Podisus nigrispinus* preying *on Alabama argillacea* caterpillars in the field. Pesquisa Agropecuaria Brasileira 37, 7-14.

Oliveira, J.E.M., Torres, J.B., Moreira, A.F.C. and Barros, R. 2002b. Effect of cotton plants and tomato plants as food supplements on the development and reproduction of the predator *Podisus nigrispinus* Dallas (Heteroptera: Pentatomidae). Neotropical Entomology 31, 101-108.

Oliveira, J.E.M., Zanuncio, J.C., Serrao, J.E. and Pereira, J.M.M. 2003. Reproductive potential of the predator *Supputius cincticeps* (Heteroptera: Pentatomidae) affected by female body weight. Acta Scientiarum: Biological Sciences 25, 49-53.

Pallini, A., Janssen, A. and Sabelis, M.W. 1998. Predators induce interspecific herbivore competition for food in refuge space. Ecology Letters 1, 171-177.

Pires, E.M., Zanuncio, J.C. and Serrao, J.E. 2011. Cannibalism of *Brontocoris tabidus* and *Podisus nigrispinus* during periods of pre-release without food or fed with *Eucalyptus cloeziana* plants. Phytoparasitica 39, 27-34.

R Development Core Team. 2010. R: A language and environment for statistical computing. R Foundation for Statistical Computing, Vienna, Austria. ISBN 3-900051-07-0, URL http://www.R-project.org.

Ramalho, F.S., Mezzomo, J.A., Lemos, W.P., Bandeira, C.M., Malaquias, J.B., Silva, J.P.S., Leite, G.L.D. and Zanuncio, J.C. 2008. Reproductive strategy of *Podisus nigrispinus* females under different feeding intervals. Phytoparasitica 36, 3037.

Relyea, R.A. 2004. Fine-tuned phenotypes: tadpole plasticity under 16 combinations of predators and competitors. Ecology 85, 172-179.

Richardson, M.L., Mitchell, R.F., Reagel, P.F. and Hanks, L.M. 2010. Causes and consequences of cannibalism in noncarnivorous insects. Annual Review of Entomology 55, 39-53.

Rickers, S. and Scheus, S. 2005. Cannibalism in *Pardosa palustris* (Araneae, Lycosidae): effects of alternative prey, habitat structure, and density. Basic and Applied Ecology 6, 471-478.

Rodrigues, A.R.S., Torres, J.B., Siqueira, H.A.A. and Teixeira, V.W. 2009. *Podisus nigrispinus* (Dallas) (Hemiptera: Pentatomidae) requires long matings for successful reproduction. Neotropical Entomology 38, 746-753.

Ruberson, J.R., Tauber, M.J. and Tauber, C.A. 1986. Plant feeding by *Podisus maculiventris* (Heteroptera: Pentatomidae): effect on survival, development, and preoviposition period. Environmental Entomology 15, 894-897.

Rudolf, V.H.W., Kamo, M. and Boots, M. 2010. Cannibals in space: the coevolution of cannibalism and dispersal in spatially structured populations. The American Naturalist 175, 513-524.

Sant'Ana, J., Bruni, R., Abdul-Baki, A.A. and Aldrich, J.R. 1997. Pheromone induced movement of nymphs of the predator *Podisus maculiventris* (Heteroptera: Pentatomidae). Biological Control 10, 123-128.

Santos, T.M., Silva, E.N. and Ramalho, F.S. 1996. Food comsumption and growth of *Podisus nigrispinus* (Dallas) on *Alabama argillacea* (Huebner) under laboratory conditions. Pesquisa Agropecuâria Brasileira 31, 699-707.

Sato, S., Shinya, K., Yasuda, H., Kindlmann, P. and Dixon, A.F.G. 2009. Effects of intra and interspecific interactions on the survival of two predatory ladybirds (Coleoptera: Coccinellidae) in relation to prey abundance. Applied Entomology and Zoology 44, 215-221.

Sato, S. and Dixon, A.F.G. 2004. Effect of intraguild predation on the survival and development of three species of aphidophagous ladybirds: consequences for invasive species. Agricultural and Forest Entomology 6, 21-24.

Schellhorn, N.A. and Andow, D.A. 1999. Cannibalism and interspecific predation: role of oviposition behavior. Ecological Applications 9, 418-428.

Schmid-Hempel, P. 2005. Evolutionary ecology of insect immune defenses. Annual Review of Entomology 50, 529-551.

Simelane, D.O., Steinkraus, D.C. and Kring, T.J. 2008. Predation rate and development of *Coccinella septempunctata* L. influenced by *Neozygites fresenii* - infected cotton aphid prey. Biological Control 44, 128-135.

Sinia, A., Roitberg, B., McGregor, R.R. and Gillespie, D.R. 2004. Prey feeding increases water stress in the omnivorous predator *Dicyphus hesperus*. Entomologia Experimentalis et Applicata 110, 243-248.

Snyder, W.E. and Ives, A.R. 2003. Interactions between specialist and generalist natural enemies: Parasitoids, predators, and pea aphid biocontrol. Ecology 84, 91-107.

Stearns, S.C. 1994. The evolution of Life Histories. Oxford: Oxford University Press. 249p.

Straub, C.S., Finke, D.L. and Snyder, W.E. 2008. Are the conservation of natural enemy biodiversity and biological control compatible goals? Biological Control 45, 225-237.

Symondson, W.O.C., Sunderland, K.D. and Greenstone, M.H. 2002. Can generalist predators be effective biocontrol agents? Annual Review of Entomology 47, 561-594.

Torres, J.B., Zanuncio, J.C. and De Oliveira, H.N. 1998: Nymphal development and adult reproduction of the stinkbug predator *Podisus nigrispinus* (Het., Pentatomidae) under fluctuating temperatures. Journal of Applied Entomology 122, 509-514.

Vivian, L.M., Torres, J.B. and Veiga, F.S.L. 2003. Development and reproduction of a predatory stinkbug, *Podisus nigrispinus*, in relation to two different prey types and environmental conditions. BioControl 48, 155-168.

Vivian, L.M., Torres, J.B., Veiga, F.S.L and Zanuncio, J.C. 2002. Predation behavior and food conversion of *Podisus nigrispinus* on the tomato moth. Pesquisa Agropecuâria Brasileira 37, 581-287.

Waddil, V. and Shepard, M. 1974: Biology of a predaceous stinkbug, *Stiretrus anchorago* (Hemiptera: Pentatomidae). The Florida Entomologist 57, 249-253.

Ware, E.R., Yguel, B. and Majerus, M. 2009. Effects of competition, cannibalism and intra-guild predation on larval development of the European coccinellid *Adalia bipunctata* and the invasive species *Harmonia axyridis.* Ecological Entomology 34, 12-19.

Zanuncio, J.C., Molina-Rugama, A.J., Serrao, J.E. and Pratissoli, D. 2001. Nymphal development and reproduction of *Podisus nigrispinus* (Heteroptera: Pentatomidae) fed with combinations of *Tenebrio molitor* (Coleoptera: Tenebrionidae) pupae and *Musca domestica* (Diptera: Muscidae) larvae. Biocontrol Science and Technology 11, 331-337.

Zanuncio, J.C., Molina-Rugama, A.J., Santos, G.P. and Ramalho, F.S. 2002. Effect of body weigth on fecundity and longevity of the stinkbug predator *Podisus rostralis.* Pesquisa Agropecuâria Brasileira 37, 1225-1230.

Zanuncio, J.C., Lacerda, M.C., Zanuncio-Junior, J.S., Zanuncio, T.V., Silva, A.M.C. and Espindula, M.C. 2004. Fertility table and rate of population growth of the predator *Supputius cincticeps* (Heteroptera: Pentatomidae) on one plant of *Eucalyptus cloeziana* in the field. Annals of Applied Biology 144, 357-361.

Zanuncio, J.C, Beserra, E.B., Molina-Rugama, A.J., Zanuncio, T.V., Pinon, T.B.M. and Maffia, V.P. 2005. Reproduction and longevity of *Supputius cincticeps* (Het.: Pentatomidae) fed with larvae of *Zophobas confusa*, *Tenebrio molitor* (Col.: Tenebrionidae) or *Musca domestica* (Dip.: Muscidae). Brazilian Archives of Biology and Technology 48, 771-777.

FINAL CONSIDERATIONS

Intraguild predation and cannibalism occurred between nymphs and adults of *P. nigrispinus* and *S. cincticeps* and helped the survival of part of the nymphs of these predators until the main prey was located.In the adult stage, this bidirectional intraguild predation and cannibalism, more frequent in *S. cincticeps*, increased the longevity of this species.

Interspecific competition did not affect the food consumption, predation rate, development and reproduction of *P. nigrispinus*. However, the nymphal development of *S. cincticeps* was shorter, with the production of larger females and lower egg viability, showing that this species is less competitive than *P. nigrispinus*. At high prey density, there was synergism in the predation rate between these predators, which is beneficial for biological control.

Podisus nigrispinus has the best potential for inundative and inoculative biological control programs, as it has a higher predation and reproduction rate. The use of this predator with *S. cincticeps* may be challenging, as the latter species is the intraguild predator and can reduce the population of *P. nigrispinus* in low prey availability, or cause it to migrate. In addition, the viability of *S. cincticeps* eggs with *P. nigrispinus* was low, showing that it would be better to invest resources in producing a greater number of *P. nigrispinus* individuals. The conservation of these species in a conservative biological control program is interesting, as they can coexist in the same environment and show synergism at high prey densities. In addition, the lower viability of *S. cincticeps* eggs in interspecific interactions with *P. nigrispinus* indicates a tendency for this interaction and intraguild predation by *S. cincticeps* to decrease.

The location of the experiment limited the natural foraging or migratory behavior of the predators. Predatory arthropods may avoid sites with other predators. In addition, more complex habitats generally contain a greater diversity of food resources and predator refuges, reducing competition, cannibalism and intraguild predation. This shows that it is necessary to develop long-term field studies to better assess the importance of these interspecific interactions between the predators *P. nigrispinus* and

S'. cincticeps.

Printed by Books on Demand GmbH, Norderstedt / Germany